SpringerBriefs in Architectural Design and Technology

Series Editor

Thomas Schröpfer, Architecture and Sustainable Design, Singapore University of Technology and Design, Singapore, Singapore

Indexed by SCOPUS

Understanding the complex relationship between design and technology is increasingly critical to the field of Architecture. The *Springer Briefs in Architectural Design and Technology* series provides accessible and comprehensive guides for all aspects of current architectural design relating to advances in technology including material science, material technology, structure and form, environmental strategies, building performance and energy, computer simulation and modeling, digital fabrication, and advanced building processes. The series features leading international experts from academia and practice who provide in-depth knowledge on all aspects of integrating architectural design with technical and environmental building solutions towards the challenges of a better world. Provocative and inspirational, each volume in the Series aims to stimulate theoretical and creative advances and question the outcome of technical innovations as well as the far-reaching social, cultural, and environmental challenges that present themselves to architectural design today. Each brief asks why things are as they are, traces the latest trends and provides penetrating, insightful and in-depth views of current topics of architectural design. *Springer Briefs in Architectural Design and Technology* provides must-have, cutting-edge content that becomes an essential reference for academics, practitioners, and students of Architecture worldwide.

More information about this series at http://www.springer.com/series/13482

Pablo Guillen · Urša Komac

City Form, Economics and Culture

For the Architecture of Public Space

 Springer

Pablo Guillen
The University of Sydney
Sydney, NSW, Australia

Urša Komac
Western Sydney University
Westmead, NSW, Australia

ISSN 2199-580X   ISSN 2199-5818   (electronic)
SpringerBriefs in Architectural Design and Technology
ISBN 978-981-15-5739-2   ISBN 978-981-15-5741-5   (eBook)
https://doi.org/10.1007/978-981-15-5741-5

This Springer imprint is published by the registered company Springer Nature Singapore Pte Ltd.
The registered company address is: 152 Beach Road, #21-01/04 Gateway East, Singapore 189721, Singapore

Contents

Chapter 1
Introduction

Abstract This is a book about how the cities utilise space and how the resulting urban form provides different ways to deal with the tangle of public goods and externalities caused by agglomeration. We rely on well-known economic thinking plus a historical analysis to why cities exist and why they have evolved to be the way they are. We identify several defining factors: the geography and the technology (both defining what is possible to do), culture (which defines what the society's goals are) and the necessary government regulation in the presence of public goods and externalities (determined both by culture and the desire to achieve positive economic outcomes). Regulation is the set of rules (not only planning codes) that underpins how markets are allowed to work in the city. Our method is also comparative as it explains the evolution of urban form in the US and how it stands in a sharp contrast with the evolution of urban form in Japan. An emphasis is put on the difference in regulations between both jurisdictions. We point out that, against the conventional wisdom, how American cities are constrained by rules that are much further from the "neoliberal" economic idea of free and competitive markets than the Japanese ones. We demonstrate how Japanese planning fosters competition and variety in the availability of goods and services. We also include an explanation of the origin of the differences in those regulations. We hypothesise how changing regulations could change the urban form to generate a greater variety of goods and to foster the access to those goods through a more equitable distribution of wealth. Critically, we point out that a desirably denser city must rely on public transport, and we also study how a less-dense city can be made to work with public transport. We conclude by claiming that changes in regulations are very unlikely to happen in the US, as it would require deep cultural changes to move from local to a more universal and less excluding public good provision.

Keywords Urban form · Regulation · Public goods · Externalities · Technology · Culture

This book is about explaining the most relevant planning and cultural differences in the way cities are allowed to function, grow and change. Those are differences in planning regimes with historical reasons we explore and rooted in culture. They reflect, but also shape, the mainstream views of the population, but have also very

P. Guillen and U. Komac, *City Form, Economics and Culture*,
SpringerBriefs in Architectural Design and Technology,
https://doi.org/10.1007/978-981-15-5741-5_1

strong economic and spatial implications. That is, most people use, see and feel the city as a consequence of planning rules they are not aware of. The aim of exploring these planning differences is not only to help to come closer to best practice given society's goals, but even perhaps allowing to direct change or bring more consistency to those goals.

Following a historical analysis of the evolution of cities influenced both by technological progress and cultural change, we argue that the unintended consequences of mass motorisation are at the origin of many, if not most, of the ills affecting contemporary cities. We propose a new approach to limit private car usage in urban environments based on the somehow *laisser-faire* Japanese experience and comparing it to the more heavy-handed, micro-managed approaches to planning used both in North America and Europe.

The car-dependent city generates problems that go much further than the obvious of pollution and excessive energy usage. First of all, universal car usage has huge implications in terms of land use, as cars need to be parked and will be actually parked most of the time. The need of parking space ("no parking no business" is, for instance, the de facto Southern California motto) means that buildings have to be spaced from each other to leave space for car parking. The result is neighbourhoods unappealing for pedestrians and leaving no option but driving. Business will be built and located to the scale and convenience of motorists. This environment makes all but impossible for small restaurants, cafes and stores to exist as they will not be even visible from zooming cars. That is, car dependency has an impact in the variety of goods and services offered. A car dependent city is therefore dominated by huge stores located far from residential zones and accessible only by car. Such city, coupled with a locally controlled strict zoning regime in terms of permitted uses, also results in a *closed* or socially exclusive city.[1] The poor won't even be visible in the rich areas and the rich have no reason to adventure driving through the poor areas as there is nothing interesting for them there. The poor in the car dependent city will suffer when their car breaks down. Soon they may not be even able to go or to look for work. As the poor and the rich live far apart, schools will be easily segregated by income and social mobility will suffer greatly. Last, but not least, there are health benefits associated to the use of public transport. Indeed, every user of public transport is a pedestrian who has to walk at least the so-called "last mile" (and probably the first too). Daily users of public transport don't need to use ridiculous walking machines in the gym or standing desks in the office as walking and standing forms part of their daily commuting to work, errands, shopping and entertainment routine.

Our main point is therefore that different planning regimes result in substantial differences in urban form, even when these differences often arise from unintended consequences of the rules chosen. For instance, the Japanese governments of the 1950s wished for a rapid motorisation but at the same time they found it unfeasible, given the narrowness of most streets, to allow on-street parking. Therefore, national

---

[1]Locally controlled planning has the aim of keeping poorer people away. Also, may be more accurate to say that a city based on public transport facilitates, but does not guarantee, openness and inclusivity. These themes will be treated in the body of the book ahead.

laws were passed by the Japanese national parliament mandating one private parking space per car and making overnight street parking illegal.[2] People in Japan did buy cars at rates not so dissimilar with their American and European counterparts but soon found them difficult to use for everyday life in the urban environment (Berri 2009).

From our reading of the architecture and urban design literature, we believe that both architects and urban designers are mostly unaware of the forces that actually shape cities and of the regulation frameworks put in place to harness and direct those forces. As a result, architects often overestimate their own contribution to the urban form. When realising they are not reaching their intended goals, they blame strawmen such as "capitalism" or "the system" and even wish for a catastrophe that would allow for a fresh start, see among a myriad of trite articles such as Aureli (2008). Quite on the contrary, we point out real world examples, mostly in Japanese and European cities, that could potentially allow architects to achieve positive outcomes in terms of more liveable cities. Japanese cities look like a straightforward result of what we believe are wise and clear planning rules. We are not saying that architecture in Japanese cities is to be emulated, as we will show examples of the commonly low-quality Japanese architecture, but we point out how the Japanese planning rules would allow for potentially excellent results. On the other hand, those excellent results would be much harder to achieve within the Anglo-American planning framework.

To us, European cities do usually look better[3] than the ones in Japan and America. We believe this is partly because the old parts of town were built before cars were available. Before the Industrial Revolution the relative cost of high-quality craftsmanship in building must have been much lower than it is now. After all, there were not that many interesting and skilled jobs for the vast majority of talented individuals before the Industrial Revolution. After all, the association of clever craftsmen is at the very origin of the Free-Mason organisation in the Middle Ages. Once education becomes increasingly available to the masses the majority of the most talented individuals prefer to become teachers, lawyers, medical doctors, professors of economics or perhaps software engineers in recent times. Craftmanship is shown in different ways, that become more rewarded by society in the form of higher salaries. Construction jobs are left to individuals who in the olden times were only deemed worth of carrying bricks to the masons. Our theory will become painfully compelling every time our estimated reader, most likely a member of the illustrated class of craftsmen in one way or another, needs anything repaired or rebuilt at home.

However, it is also true that unlike in Japan European buildings have better survived the pass of time and the destruction of war.[4] In any case, cities in Europe are nowadays much more car dependent and crowded with cars than their Japanese

---

[2]On street parking is seldom allowed. When allowed, is often metered and indeed, cars still parked after midnight are towed, see Barter (2014).

[3]At least the areas frequented by tourists. Admittedly a value judgement.

[4]The widespread use of timber in Japan even for large and symbolic buildings, such as temples and palaces, has a lot to do with that. European medieval cities were also mostly built with timber, besides the cathedral and the castle. When the city burned, new buildings were made of more durable materials. Japanese timber construction was prevalent up until mid twentieth century. Buildings

counterparts. Whenever a city in Europe has managed to limit car usage, that has happened as a result of a very micro-managed approach and, sometimes, a substantial cost in terms of public transport subsidy.

In summary, we are advocating for an urban form that is sufficiently dense to foster encounters at a human scale, encourages the supply of variety goods and services, provides opportunities for recreation and the enhancement of the soul, it is not planned around the idea of exclusion and it is serviced by affordable and efficient public transport that keeps pollution to a bearable minimum. It seems that such a city would attract and shape the best minds to generate wealth which, appropriately taxed, could provide and expand the public goods and thus grow in a virtuous cycle.

References

Aureli, P. V. (2008). *The project of autonomy: politics and architecture within and against capitalism* (Vol. 4). New York: Princeton Architectural Press.

Barter, P. (2014). *Japan's proof-of-parking rule has an essential twin policy. Reinventing parking.* https://www.reinventingparking.org/2014/06/japans-proof-of-parking-rule-has.html. Last Retrieved on March 01, 2020.

Berri, A. (2009). A cross-country comparison of household, car ownership: A cohort analysis. *IATSS Research, 33*(2), 21–38.

Sorensen, A. (2005). *The making of urban Japan: Cities and planning from Edo to the twenty first century.* Routledge.

---

destroyed by fire-bombing during the war were replaced in haste during the post-war economic recovery resulting in not-so-pleasant looks. See Sorensen (2005).

Chapter 2
Why Cities Exist?

Abstract We argue cities exist are the result of economics forces of agglomeration mediated by technological progress. That is, urban growth is fuelled by economically advantageous division of labour. Available technology is the most important constraint to city growth. However, other factors such as political stability, peace and the control of plagues are also important.

Keywords Division of labour · Specialisation · Trade · Agglomeration

Humans become settled in a particular place as a result of the Neolithic revolution. That is, the discovery of agriculture and husbandry, which tie people to the land. Initially a group of humans with family ties would settle in a piece of land and work together to produce enough food for the clan. In some places, such as particularly fertile plain of Mesopotamia and the Nile valley, agriculture is especially productive.[1] There, a relatively small group of farmers was able to produce enough food for many people, much more than was needed for themselves and their families. Surplus appeared, so it was no longer necessary for the entire workforce to toil the land.

So, some people will specialise in farming, but others can now focus on different occupations and thus become full time builders, craftsmen, soldiers, priests or artists. This division of labour encourages specialisation. Specialists will produce goods or services they don't need for themselves and trade with other specialists. Because of trade everyone can now be better off than in autarchy, a situation in which people only produce for themselves. Another interesting feature of production technologies is that often specialists can produce better and more when they are next to each other and, critically, also when they are close to people interested in their trades. That is, close to the market. Those are the forces fuelling the process of agglomeration. Cities are an economic consequence of specialisation, trade and agglomeration.

Therefore, cities would thrive as much as they can, sustain their population and attract new dwellers, brought both by the economic opportunity and the fascination by the plethora of opportunities the city provides. That is, not the variety of goods provided but also a diverse, large population who may be willing to try new and

---

[1] Most of our historical claims are standard and can be checked in any universal history manual, see for instance Gombrich (2005) for a fairly comprehensive world history up to the twentieth century.

P. Guillen and U. Komac, *City Form, Economics and Culture*,
SpringerBriefs in Architectural Design and Technology,
https://doi.org/10.1007/978-981-15-5741-5_2

different things. Architects, artists and professors come to the city because that's where their abilities are appreciated and paid for. All the wonders the city has to offer, its arcades and plazas, its culture, its sophisticated food and theatre cannot exist if architects, artists and professors, chefs and actors are not paid. If their job is better in Melbourne, they might just leave Sydney.

Another important fact about cities is that they cannot be taken for granted. For instance, they came next to disappearing in Western Europe at the time of the barbarian invasions. That happened not only because cities where directly attacked and sacked by invaders but also because, once the authority of the Roman Empire of the West collapsed, nothing could keep slaves working the fields thus agricultural surplus disappeared. War and bandits cut trade routes. People had no option but to go back to the fields to avoid starvation. Without cities, art and culture soon stagnated in Western Europe, but flourished in the East where law and order still prevailed. For several centuries to come, Byzantium was the new Rome.

Technology has a big role in increasing the productivity of agriculture and therefore causing migrations to cities. The iron Roman plough created an empire. The steel plough of the early XIX century pushed again masses of workers from the fields to the factories. The green revolution epitomised by tractors, chemical fertilisers and insecticide finished the job in the XX century. It is worth noting that many of those migrants were not only attracted by the opportunity the city had to offer but somehow expelled from their traditional occupations in the fields. Many were escaping poverty but ended up in a poor city slum.

Reference

Gombrich, E. H. (2005). *A little history of the world*. Yale University Press.

Chapter 3
Cities Are More Important Than Ever

Abstract We show how, contrary to predications made in the twentieth century, the advances of transportation and information technology have not slowed down the forces of agglomeration. On the contrary, because of the increased human and physical capital accumulation plus the availability of desirable goods and opportunities, city growth has been accelerating. The world is going through a gradual but seemingly unstoppable process of urbanisation.

Keywords Economies of scale · Capital accumulation · Contemporary urbanisation

Not that long ago many scholars have expressed doubts and hopes about the future of the city. Telecommunication technology and motorisation were seen by most as ways to stop the forces of agglomeration and allow humans to go back to live close to nature. Congested, polluted cities were hoped to be a thing of the past by the twenty-first century. Those hopes have been dashed. A recent study shows[1] how cities of different eras aren't as different as we might think. Modern settlements grow similarly to their ancient counterparts. In particular, city growth in all ages is characterised by productivity increasing faster than population. The city is a source of economies of scale.[2] By and large, the economic success of the city is the main driver of population growth and urbanisation. Note that the phrase "economic success" has to be understood in a wide sense. On one hand cities are a good place for production given the economies of scale fuelled by specialisation, but critically and often overlooked, cities are also a good place for *consumption* as they offer a plethora of varied goods, services and other opportunities[3] not available elsewhere. That explains why not everyone moves to Dubai, which offers excellent salaries (related

---

[1] Ortman et al. (2015).

[2] That is productivity, or output per worker, increases with the scale of production, see for instance Nicholson and Snyder (2015) for a simple yet accurate explanation of most of the microeconomic terms used in this book.

[3] Basically, related to cities being full of people you don't know yet and thus, as a Chicago School economist would put it, enhancing the "choice set" in many realms, i.e. sexual partners.

P. Guillen and U. Komac, *City Form, Economics and Culture*,
SpringerBriefs in Architectural Design and Technology,
https://doi.org/10.1007/978-981-15-5741-5_3

to production), but comes with serious shortcomings in terms of goods, services and other opportunities on offer.[4]

The forces of specialisation and agglomeration are nowadays stronger than ever before. As a result, population is concentrating in cities faster than at any time in human history, Ritchie and Roser (2018). The world is going through a gradual but seemingly unstoppable process of urbanisation. If by the beginning of the twentieth century about 15% of the world population lived in cities, this proportion increased to 50% in 2007. The accelerated shift of population from rural to urban areas has also been accompanied by a very strong population and economic growth. World population went from 1.6 billion in 1900 to 6 billion in 1999.[5] All in all, urban population went from about 250 million at the beginning of the twentieth century to 3 billion at the end of the century. That's a 12-fold increase or a 1100% increase in percentage terms. Urbanisation and population growth are a staggering, unprecedented phenomena in human history. Both processes are a result of technological and cultural changes that, starting in the mid-eighteenth century, gave birth to the very efficient although still evolving form of production known as capitalism. That is a mode of production characterised by capital accumulation. In the pursue of ever higher profits, current profits are invested to expand the production capabilities by purchasing new and/or better machines (capital) or more sophisticated and efficient ways of combining capital with labour.[6] Indeed, for good and bad, this is main force behind our thriving, growing cities.

One fact common to all contemporary cities is the huge impact motorisation has in its organisation. Humans now mostly live in cities, but most of the newer ones have been built to move around in cars. The older ones had to accommodate to the new technology. Far from being back to nature humans now live in a tar and steel jungle where the rich can afford a lawn and a pool as, maybe, a poor substitute of a meadow and a river of clear waters. Many people rub bumpers rather than shoulders. The rush hour is still alive and well.

---

[4]That is, if you don't quite like golden taps and air-conditioned beaches.

[5]This fast increase in population was much unexpected by the average person in the mid-twentieth century. For instance, in the 1940 s Isaac Asimov assumes fairly low overall populations in his futuristic science fiction novels, see Asimov (2004).

[6]Note that capital accumulation is not unique to countries commonly known as capitalists. Both capital accumulation and technological progress were indeed at the core of the planned economies in so-called socialist countries. The essential difference lies in that most of the investment is decided centrally in a planned economy rather than decided by privately owned companies in a not-so-planned economy. Public infrastructure and public goods are still provided by the state even in the most capitalist economies. The urban governance problem, particularly related to the city, lies on what infrastructure to build and which public goods to provide for the city in order to reach which goals. These problems will be discussed in more detail later on in the book.

References

Asimov, I. (2004). *I, robot*. Spectra.

Nicholson, W., & Snyder, C. M. (2015). *Intermediate microeconomics and its application, twelfth edition*. Cengage.

Ortman, S. G., Cabaniss, A. H., Sturm, J. O., & Bettencourt, L. M. (2015). Settlement scaling and increasing returns in an ancient society. *Science Advances, 1*(1), e1400066.

Ritchie, H., & Roser, M. (2018). Urbanization. *Our World in Data*.

Chapter 4
Public Goods, Externalities and the City

Abstract We explain the concept of public goods as understood in economics. That is, non-rivalrous and non-excludable goods as opposed to rivalrous and excludable private goods. We also explain the concept of externality as the effect on society as a whole. We show how markets cannot effectively provide neither public goods nor goods that involve externalities. We argue that cities, as a tangle of public goods and externalities, need effective governance and thus regulation.

Keywords Public goods · Externalities · Non-market provision

As normally understood public goods are commodities or services provided without profit, even for free or a nominal fee usually by the government or by a private organisation on behalf of the government.

We will use in this book a somehow more sophisticated definition of public goods borrowed from economics. Public goods are thus characterised as being both *non-excludable* and *non-rivalrous*, see Nicholson and Snyder (2015). A good is non-excludable when it is not possible to prevent people who did not pay from having it. Streetlights are clearly non-excludable; anyone walking down the street at night may enjoy them. The safety provided by a well-functioning police force is also non-excludable. A good is non-rivalrous if it can be used by more than one person at the same time in a way such that consumption from one individual does not detract from the enjoyment of other individuals. Streetlights are also non-rivalrous, they can be enjoyed by many people at the same time. Again, the same can be said about the safety provided by a well-functioning police force.[1] Note, however, that when a constant number of policemen need to look after an increasingly large group of people safety can be compromised. The good becomes *saturated*, rivalry comes into place and we don't have a public good anymore.

---

[1]In fact, the legal system as a whole is a public good. Without a legal system the enforcement of property rights would be impossible. The mere existence of markets therefore relies on a public good that must be provided by the government. This fact is as essential as overlooked.

© The Author(s), under exclusive license to Springer Nature Singapore Pte Ltd., part of Springer Nature 2020
P. Guillen and U. Komac, *City Form, Economics and Culture*,
SpringerBriefs in Architectural Design and Technology,
https://doi.org/10.1007/978-981-15-5741-5_4

Private goods, such as pears and laptops, are obviously both excludable and rival-rous. Goods that are non-excludable and rivalrous, such as a public good that became saturated, are called "common pool resources", i.e. fish stocks in the oceans. Finally, some goods are excludable and non-rivalrous like a movie shown in a cinema. Those are, by the way, called "club goods" in economics. For instance, a non-toll unsaturated road can be understood as a public good. Once it becomes saturated it is a common-pool resource.

Now it is useful to argue that non-excludable goods are unlikely to be provided by a for-profit private entity. For instance, if a fireworks show can be seen from people's balconies, not many people would be willing to pay for it. Even if most people who like fireworks were willing to pay $10 for a show, many (or most) could not be compelled to do so. Therefore, a private fireworks show would not happen because it is unlikely to be a good business. That is, free markets will not provide public goods. Free markets are also bad at exploiting common-pool resources.[2] Club goods can be efficiently provided by the market under certain conditions too technical to discuss here.

A concept close to public goods is that of externalities. A positive externality arises when something someone affects positively somebody else who does not pay for it. That is, a $1000 firework show payed for by a rich die-hard fireworks enthusiast entails a positive externality for everyone else who enjoys it but does not pay for it. Die-hard rick fireworks enthusiasts, willing to foot the bill all by themselves, are a rare species so fireworks are most of time payed by the city government and ultimately funded by taxes.

Negative externalities can be thought to be linked to *public bads*, that is non-rivalrous and non-excludable things *everyone* dislikes.[3] For instance, something someone does for private profit negatively affecting other people who did not pay for it. That could be the case, for instance, of driving in a congested road. More to the point, driving to work produces a private benefit and several negative externalities. Pollution and congestion are the two most obvious. We will argue afterwards that planning rules that encourage or impose car dependency in a large urban area generates other, perhaps more pervasive, negative externalities in terms of land use and city form. Also, it is very important to understand that free markets are not a good way of dealing with externalities. As with public goods, a perfectly competitive industries will produce too much of a negative externality and too little of a positive externality.

Finally, note that public goods and externalities need to be considered relative to location. For instance, CO_2 emissions entail a global negative externality in terms of climate change. On the other extreme, local planners may put limits to low socio-economic families to settle in their municipality as those would bring a negative

---

[2]For instance, an unregulated, perfectly competitive fishing industry would result in depletion of the fish stocks, see Nicholson and Snyder (2015).

[3]Note that something can be a *good* for some and a *bad* for others. Fireworks are a good example.

externality in terms of lowering real estate values, decreasing the quality and increasing the cost of public goods provided locally etc. Similarly, Euclidean zoning[4] limits density and imposes single use zones to ameliorate the negative externality caused by traffic in residential areas. We will argue that these forms of planning, while being locally effective have a negative regional or nationwide effect. That is, forcing the poor to live next to the other poor creates huge negative externalities in terms education and crime outcomes that affect the city, region or nation as a whole. Limiting density and separating zones by use has the effect of increasing overall traffic and congestion. It just pushes it away from particular, often affluent, residential areas.

We have argued that cities are growing fast because of their increased capacity to produce wealth. That is now mostly happening in the form of highly valuable services. Cities, however, entail a huge tangle of non-private goods (public, common pool and club) and a variety of positive and negative externalities that *must* be dealt with by government intervention, provision or regulation.[5] For instance, public space generates several public goods *at the same time*. Architects and city planners need to be well aware of how the design of the city affects in a positive or negative way the provision of those goods.

It is also useful to differentiate between *a* public good and *the* public good. The former is a non-excludable, non-rivalrous good and the latter what is good, in the sense of positive, for the public in general. It could be said that public goods are provided for the public good, the benefit of the public. And that should indeed be the goal of government: nothing else other than the public good.

References

Fluck, T. A. (1986). Euclid v. Ambler: A retrospective. *Journal of the American Planning Association, 52*(3), 326–337.
Nicholson, W., & Snyder, C. M. (2015). *Intermediate microeconomics and its application, twelfth edition*. Cengage.

---

[4]Euclidian zoning, Fluck (1986), will be explained in detail in Chap. 8.

[5]Some economic thinking suggests that the cost of market failure due to public goods, externalities etc. is actually lower than the cost of government failure. That is the cost imposed on society in terms of taxes, lobbying, corruption and so on. Notwithstanding government ought to be less than perfect, we deeply disagree with this line of thought.

Chapter 5
Governing for the Public Good: The Problem of City Governance and Planning

Abstract We discuss the problem of city governance in general and with regards to urban planning in particular. Although cities exist because of their capacity to generate wealth, we do not believe that elected public officials should focus solely on the maximisation of economic growth. Indeed, cities are not only centres of production but also residence and consumption of private and public goods. Elected official should then strive to maximise a social welfare outcome rather than a merely monetary one. Any planning policy is a form of government intervention or regulation. Given the complexity of interconnected public goods and externalities posed by agglomeration, the need for regulation is unavoidable.

Keywords City governance · Public good · Urban planning · Regulation

We have argued that specialisation and trade foster agglomeration. Cities grow because they are a hotbed of economic opportunity. Should city governance be thus focused on the generation of wealth? Of course not, good governance is to be focused on the public good, which is what economists call maximising social welfare. Of course, that does not simply imply maximising the generation of wealth. Politicians would ideally have some abstract and overarching goals, for instance equality of opportunity, the provision of certain public goods, a certain degree of redistribution of income and economic growth. Those goals would together generate a particular social welfare outcome. Different political platforms would emphasise different aspects in terms of social welfare. Some would insist on income distribution while others would support economic growth combined or not with universal education as a mean to achieve equal opportunity as the ultimate goals of society. In a democratic system people choose a political platform through voting to organise society according to a particular set of principles, for a limited time.

A politician likely to become a planning minister should seek advice on how to reflect their ideals in the planning portfolio. In the best-case scenario, the job of such politician is to convince a majority of the electorate of the merits of a political platform on planning. We are not saying that a politician should ignore any ideas or suggestions coming from the public, but they should see how they fit with expert advice and the political principles. For instance, if one asks the public about placing

P. Guillen and U. Komac, *City Form, Economics and Culture*, SpringerBriefs in Architectural Design and Technology, https://doi.org/10.1007/978-981-15-5741-5_5

15

strong limitations on street parking, a vast majority would be initially against. A good politician should be able to convince the public of the merits of such a proposal. This book aims to provide good reasons for this and other policies concerning public goods and externalities in the urban environment.

Any planning policy is a form of government intervention or regulation. Given the complexity of interconnected public goods and externalities posed by agglomeration, the need for regulation is unavoidable. It is rather a question of which set of rules are best suited to achieve a particular outcome that cannot be reached by market forces alone.[1] We will advocate for less, simpler, easier to enforce planning controls. Curiously enough, planning in jurisdictions generally understood as more pro-market and utterly *neoliberal*, such as the US and to a lesser extent the UK and Australia, usually have more and more inflexible planning rules that require a lot of micromanagement and generate boring, car-congested cities which may not even be the best for wealth generation. However, there is nothing essentially pro-market in the strict zoning regimes characteristic of most of the US. If anything, this planning approach has more to do with a planned economy than with the free market. We find that a fascinating and very interesting contradiction. This could be understood by taking into account that the strict American planning regime tends to generate local monopolies (and exclude the undesirable poor from high quality local public goods). Indeed, in a low density, zoned environment there would be just one shop of one kind in each neighbourhood. That's even the case by design in shopping centres contractually limiting the number of shops of the same kind that are admissible under the same roof. That's far from a perfectly competitive market, but a regime that enacts unsurmountable barriers to entry by making land unavailable to competitors. That is, a free market, perhaps, but only for the incumbent and definitely not perfectly competitive.[2]

Reference

Nicholson, W., & Snyder, C. M. (2015). *Intermediate microeconomics and its application, twelfth edition*. Cengage.

---

[1] Market forces could help to achieve some outcomes if they are properly channelled by regulation.

[2] In microeconomic theory a perfectly competitive market achieves full efficiency in the absence of market failures such as public goods or externalities. Perfect competition also needs to assume free entry and exit of firms, Nicholson and Snyder (2015). That is, a free market may be far from perfectly competitive. Sometimes governments intervene to push markets closer to perfect competition (i.e. antitrust laws).

Chapter 6
Growth and Shape of the Pre-industrial City

Abstract We analyse the growth and shape of the pre-industrial city as a result of the transportation technology available before the mechanisation of transport. Such city is constrained in size by walking speed. Because of the need of minimising transportation time or cost it has, necessarily, one centre and is fairly dense. The location of pre-industrial cities was also often determined by access to water-based transportation. We point out to New York and Venice as two examples of cities already preeminent before the mechanisation of transport. New York adapted to the new technology, but that is not the case for Venice.

Keywords Pre-industrial growth · Transportation · Technology

We have so far discussed *what to do*, goals that are based on preferences or, in other words, culture. What *is possible to do* is in the realm of technology.

For millennia, the growth and shape of cities has been constrained by the transportation technology available.[1] For a long time, nothing could move faster on land than a horse. Most people run their errands by foot. If we think of commutes of up to one hour[2] as common, we obtain the maximum size of a walking-based city. The outer suburbs cannot possibly be further than 3 km from the centre. For the same reasons the old slow speed city would also need to be rather dense. Further than that, the speed constraint results in a city with only one centre combining as many functions as possible as travel between different centres would be too onerous in terms of time.[3]

Transportation technology helps to explain not only the shape of old cities but also their geographical location. For instance, prior to railways any sizeable city in the US was located either on the coast or at a major inland waterway, as water-based transport was at the time the cheapest and most efficient way to move freight around. Many of those cities are still important, just think about New York, Philadelphia,

---

[1] And of course, also by the building technologies available.

[2] The so-called Marchetti's constant: the total time spend in commuting is stable along history and around the world and equal to one hour per day, Marchetti (1994).

[3] Although the king may like to live in Versailles.

© The Author(s), under exclusive license to Springer Nature Singapore Pte Ltd., part of Springer Nature 2020
P. Guillen and U. Komac, *City Form, Economics and Culture*,
SpringerBriefs in Architectural Design and Technology,
https://doi.org/10.1007/978-981-15-5741-5_6

Chicago or Boston. Being large and economically powerful they had the resources to invest in adapting to new transportation technologies. The initial success of New York City was based on Manhattan being an island between a river (the Hudson) and sea channel (the East River). The Hudson offering excellent communication with the hinterland. A huge amount of resources needed to be spent first on adapting the city to rail transportation, access to motorways and air transportation. Some very old cities were somehow easy to adapt to car usage. Rome, for instance, is full of cars, and one of the most polluted and congested cities in Europe, but it is still a thriving metropolis and the capital of Italy, the 8th largest economy in the world. In the same country, we can think of Venice as an extreme case in failing to adapt to changing technology. The success of the Venice Republic, a leading a trading empire that dominated the east Mediterranean for centuries, was based on easy access to the sea. However, the very reasons of Venice's success were the seeds of decadence. Unlike Manhattan, Venice is not one island, but a collection of swampy islets separated by canals. Venice was largely unable to adapt to motorised land transport, not only automobiles but also trams and railways. Some artists were able to see the problem with clarity. The Italian Futurists, led by the poet Marinetti, were enraged by the obsolescence of Venice and the general will to keep it as it was. They proposed to dry up the canals and fill them up with the rubble taken from crumbling *palazzi*, the Canal Grande should be dredged and widened to become a busy commercial port.[4]

The Futurists eventually got a little bit of what they wanted. The 3.8 km Ponte Littorio causeway, opened by Mussolini in 1933 links Piazzale Roma to the mainland. Nowadays Venice's road and rail connections mostly bring in tourists rather than raw materials or commuters. The same is true for Venice's harbour, now dominated by cruise ships. Venice has all but lost its past importance as a commercial, industrial and cultural hub.[5] Note that other metropolis on the sea faced, to a lesser extent a similar challenge but were more or less successful to adapt to the new technological conditions.

References

Marchetti, C. (1994). Anthropological invariants in travel behavior. *Technological Forecasting and Social Change, 47*(1).

Rainey, L., Poggi, C., Wittman, L. (Eds.) (2009). *Futurism: An anthology*. Yale University Press.

---

[4]Marinetti's *Futurist Speech to the Venetians* can be found in Rainey et al. (2009).

[5]Although Mestre and Marghera, across the Venice lagoon, are thriving industrial centres belonging to the Venetian metropolitan area, Venice lacks a business centre and depends almost exclusively on tourism for subsistence.

Chapter 7
The Raise of the Rail-Based Mechanical City

> *[...] in New York, we speak within limits when we say that a lady not unfrequently is compelled to wait half an hour" to cross the street, "and even then she makes the crossing at any point below the Park at her peril.*
> Harper's New Monthly Magazine, Volume 9, 1854.

Abstract We study the effect of mechanised rail-based transport in the shape and growth pattern of the city during the early Industrial Revolution. Often the railway required substantial changes such as the demolition of parts or all the city walls. However, the most enduring effect on city shape came from the use of railways and tramways for transportation within the city and its suburbs. Although still having one centre, this became larger and denser. On the other hand, residential communities accessible by rail grew around the city proper, London's Metro Land being a paradigmatic example. Nevertheless, people still needed to walk the last mile and goods moved by horse carts determining the location of commerce and industry not far or mixed with dense residential areas. Increasing congestion, a public bad and a by-product of strong economic growth, could not be overcome until the adoption of electric underground city railways.

Keywords Industrial revolution · Rail transportation · Technology · Congestion · Public bad

The industrial revolution brought mechanised manufacturing and transportation. Railways and trams together with industry resulted in strong economic growth affecting mostly the city. Initially railways were built to link inland cities with ports, provincial cities with capitals and mines with factories. The railway station was located as close as practical to the old administrative and commercial centre. Often the railway required substantial changes such as the demolition of parts or all the city walls, in any case rendered unpractical by the development of modern artillery. In any case, the adoption of the new transport technology changed the shape of the city even in its initial stages. Very soon it became clear that railways would not only be useful to link cities with each other but to provide urban transportation. Soon after railways became

P. Guillen and U. Komac, *City Form, Economics and Culture*,
SpringerBriefs in Architectural Design and Technology,
https://doi.org/10.1007/978-981-15-5741-5_7

practical, short lines using steam trains with locomotives and carriages started to fan out from inner city terminals in London and Paris.[1]

The new urban networks were shaped radially, linking the old centre to new areas of expansion or nearby towns, such as Greenwich and Versailles, soon to be engorged by the metropolis. Street running light railways known as trams or streetcars added a cheaper option that could use the existing road system and reach many more places. Trams and railways made possible a city much larger, both in physical size and population, than what blood traction ever allowed for. At the same time, and thanks to the industrial revolution also propelled by steam and the increased productivity of agriculture[2], strong economic growth and immigration from the countryside made a bigger city necessary. The city centre grew fast and became denser.[3] Note that at this point in time trams and railways allowed for faster, higher capacity transportation but they both run on rails that are expensive to lay. Trains and trams run therefore in a determined route and according to a timetable. The city can therefore expand around stations and tram routes. Because railways originate near the old centre this becomes bigger and stronger. The 19th century rail-based city has still one centre, but outlying rail served residential suburbs are now possible.

In such a city people still need to walk the last mile to their occupations. Often walking within the city centre may well be more efficient than using public transport. However, freight has no legs. Goods still needed to be carried from stations to their final destinations in carts. A central location is good for business, so with the help of new construction techniques using steel beams city centres just became denser and roads soon become too clogged with carts, pedestrians and horse or electric trams. Large scale congestion appeared and came to any successful city to stay. Congestion is inconvenient and causes delays. Nevertheless, the most important problem posed by congestion is it does not allow the centre to grow any further and thus hinders economic growth. It did not take long to find a way to pack more people and business in the city centres. The solution was found in the extremely congested London: the underground railways. The London Underground started from railway companies

---

[1] The London and Greenwich Railway reaches the latter as early as 1838. In Paris, a railway between what is today Gare Montparnasse to Versailles opened in 1840. Both railways were built and run as short suburban lines. The LGR was eventually prolonged, but what became to be called "Ligne des Invalides" never went pass Versailles and is today a branch of the commuter railway known as RER C. *Petit Ceinture to Auteuil.*

[2] See Kriedte (1983).

[3] We are focusing on the effects of mechanised transport in the increase of productivity and city growth. We should, however, acknowledge the great importance of sanitation in allowing for city growth. The Cloaca Maxima in Rome was a massive underground sewer and one of the most critical pieces of infrastructure in the antiquity. Similarly, fast growing cities in 19th century western Europe had to solve the sanitation problem. Victor Considerant, a social reformer, wrote in 1845 *"Paris is an immense workshop of putrefaction, where misery, pestilence and sickness work in concert, where sunlight and air rarely penetrate. Paris is a terrible place where plants shrivel and perish, and where, of seven small infants, four die during the course of the year."* He was not the first, according to Voltaire French cities were *"established in narrow streets, showing off their filthiness, spreading infection and causing continuing disorders."* Sanitation, however, is not a contentious topic so we will not discuss it much further in this book.

willing to connect their inner-city terminals in an efficient way. The concept of an urban underground railway succeeded, and soon other companies started to build their own lines. The electrification of those railways, which started again in London, and in different ways also in Budapest and Boston, made them much more practical and popular. Electrification also made possible to bore deeper, small diameter tunnels.[4] Paris followed suit, but unlike London, from the very beginning the city wanted a comprehensive network to cover the whole city. That was opposed by the French government, which wanted an urban railway capable of carrying main line trains, so a standard track gauge was imposed. The city circumvented this ruling by choosing small tunnels, unsuitable for main line trains. Curiously a similar dispute took place in Vienna at pretty much the same time. In that case the government and the military imposed steam traction in the newly built Stadtbahn network.[5]

In any case, urban underground railways allowed for higher population and business density in the city centres. Around the same time tramlines were also electrified.[6] An explosion of tram and rail-based suburbs appeared around the city. They promised comfortable accommodation in detached or semi-detached houses for the middle and upper-middle classes. Tram based suburbs were located relatively close to the city centre and sometimes were absorbed by it. Commuter rail towns where built along train lines, surrounding older villages and towns. Many examples can be found in the so-called home counties surrounding London. Some early garden cities started as rail-based towns in which people would walk to the station to catch a train to London. Although garden city theoreticians sought to build self-sustaining communities providing most of the employment locally that did not quite work as planned. For instance, Letchworth was conveniently placed on the Great Northern Railway at about 55 km from London King's Cross station. Sometimes railway companies also owned and developed the land around railway stations. That is the case of the London Metropolitan Railway's suburban development to the North of London known as Metro-land. This is an exception rather than the norm in the United Kingdom. Indeed, unlike other British private railways the Metropolitan Railway was allowed to retain surplus land, that is, land required for building the rail line but not necessary for the operation of the railway. This land was managed by the nominally independent Metropolitan Railway Country Estates Limited (MRCE), although effectively controlled by the Met's directors. Metro-land was the brand with which the MRCE promoted the dream of a modern home surrounded by beautiful countryside and linked to London by fast electric trains (see Fig. 7.1). The Metropolitan Railway morphed into the Metropolitan Line of the London Underground. Metro-land, a string of suburbs in Buckinghamshire, Hertfordshire and Middlesex, is a desirable

---

[4]To this day the London Underground consists of relatively wide subsurface tunnels built originally for steam trains (i.e. the Circle, District and Metropolitan lines) and the deep tube lines used by trains with a distinctively small cross-section (*loading gauge* in railway lingo).

[5]The Wiener Stadtbahn was eventually rebuilt and electrified in the 1920s and evolved to be integrated either in the Vienna Metro or the commuter rail S-Bahn.

[6]The first practical tramway electrification designed and executed by Frank Sprague's firm for the Richmond Union Passenger Railway in Richmond, Virginia. Boston, Massachusetts and many other cities across North America and Europe followed suit.

Fig. 7.1 Metro-land promotional poster. *Source* Wikipedia

Fig. 7.2 Pinner high street, from Wikimedia commons

and expensive place to live that has long shaken off its aspirational hue and grew some patina. It still takes just 25 min from Pinner (Fig. 7.2) to Baker Street, try to beat that by car in London. Trains depart every 6 min.

The Pacific Electric system from Los Angeles, the famous Red Cars,[7] was built to serve suburbs along the line. There was even less doubt in this case of the connection between the owners of the Pacific Electric, Henry Huntington and the Southern Pacific Railroad, and the real estate business. The Red Cars were indeed run at a loss for most of the time and used to serve the needs of newly developed suburbs built along the lines. Unlike London's Metropolitan Railway, most of the Pacific Electric tracks were often embedded on roads, so the Red Cars became to be seen as a nuisance for motorists as soon as car ownership became common in the 1920s. That, together with the unprofitability of the business rushed its complete closure in the 1950s.

---

[7]The Yellow Cars of the Los Angeles Railway covered the inner city and moved many more people than the Red Cars, but they are far less famous. Actually, despite all the hype, the Pacific Electric patronage was rather underwhelming and always below that of the Yellow Cars, see Stargel and Stargel (2009). The busiest line of the Pacific Electric, the Sawtelle to Santa Monica, moved a bit over 2.5 million passengers in 1929, its best year.

Development models similar to Metro-land also came to existence in Japan during the first half of the 20th century, initially by private railways in the Kansai area comprising Osaka, Kyoto and Kobe. The trunk railway lines in Japan were mostly built by private companies during the second half of the 19th century. The Japanese government, in an attempt to speed up railway construction, nationalised the trunk intercity railways in 1906, Ike (1955). That left private companies almost out of business. Only local lines, many times little more than modest tram lines[8] linking small towns in the countryside with a city, were not nationalised. Private companies were also allowed to newly build railways of lesser importance. Among those there were the Hanshin main line, the very first Japanese interurban line, and the Minoo Arima Electric Railway which soon became the Hankyu railway. Ichizo Kobayachi, Hankyu's owner, had to compete with both the National Railway and the Hanshin line for the Osaka to Kobe market. Being the newcomer Hankyu could only be built in the hilly, less populated areas to the north. Land there, however, was relatively cheap so Kobayachi invested in real estate. Like before in London, the 1920s Kansai middle class salaryman could now move to a relatively large, for Japanese standards, comfortable and reasonably priced house near the countryside. Hankyu made money by selling those houses, which payed for the construction of the railway. Every time a house was occupied there was a family of new customers using the railway for commuting and shopping, see Doi and Kawaushi (1995). That payed for the day to day cost of the railway and allowed, and in a sharp contrast with the long-gone Pacific Electric still allows to this day, Hankyu to be run at a profit. To keep business up, Hankyu opened a theatre at one end of a branch line, in Takarazuka, and a department store near Osaka-Namba terminus.

This model of suburban development was followed by many other private railways in many Japanese cities, like the Tokyu in the Kanto area or the Meitetsu in Nagoya. To this day many Japanese department store chains are still own by private railway companies with their flagship store next to the city terminal. Japanese private railways are among the largest and most prestigious companies in the country. They usually lead a group of companies comprising a variety of businesses.

References

Doi, T., & Kawauchi, A. (1995). A historical viewpoints of image formation of the suburb developed by Hankyu railway. *Historical Studies in Civil Engineering, 15,* 1–13.
Ike, N. (1955). The pattern of railway development in Japan. *The Journal of Asian Studies, 14*(2), 217–229.

---

[8]These electric railways were known as "interurbans" in the US. They served not-so-densely populated areas around cities or linked rural communities with main line railway lines. Interurbans used the electrification technology of trams, but interurban cars or trains were heavier and faster. Interurban cars usually shared tram tracks to reach the city centre. They mostly disappeared in the US after motorisation. The South Shore Line is one of a few surviving American interurbans. The already mentioned Pacific Electric railway run an extensive interurban network around Los Angeles until its demise in the 1950s.

Kriedte, P. (1983). *Peasants, landlords, and merchant capitalists: Europe and the world economy, 1500–1800*. Leamington Spa, Warwickshire: Berg.
Stargel, C., Stargel, S. (2009). *Early Downtown Los Angeles*. Arcadia Publishing.

Chapter 8
Motorisation and the City: America Leads the World

Abstract We study the causes and effects of mass motorisation in the United States of America in terms of city shape and function. First, we build up the historical context that explains how the motor car was adopted by the masses in the 1920s. Then, we study the evolution of the planning regime necessary to accommodate the motor car into the fabric of American cities throughout the twentieth century. We argue that such planning regime, characterised by complex and micromanaged ordinances imposing strict zoning, low density and minimum parking requirements, is far from the American free-market ideals. It is, however, a way to manage some externalities caused by car dependency and fits with cultural norms regarding local provision and funding of public goods. We emphasise the problematic spatial aspect of car dependency as a result of American planning. We also analyse the role of racial segregation and exclusion in the strong political choice in favour of motorisation. Finally, we discuss several contemporary schools of thought and concepts arising from the American approach to planning.

Keywords Euclidean planning · Minimum parking requirements · Local public goods · Edge city · Jane Jacobs · Urban economics · Racial segregation · Gentrification · Planned pedestrian hostility

We are providing a relatively long explanation of motorisation in America as it yields a blueprint for the external effects of private car usage and offers an interesting example of how to deal with them. There are some particular American traits, i.e. the importance of racial discrimination, the availability of cheap land and the prevalence of bottom-up planning, which we believe explain the differences between America and other planning regimes.

P. Guillen and U. Komac, *City Form, Economics and Culture*,
SpringerBriefs in Architectural Design and Technology,
https://doi.org/10.1007/978-981-15-5741-5_8

By the early twentieth century metropolitan cities had a large, dense urban core served by underground[1] railways and tramlines. Cities also had less dense suburbs connected to the centre by tram or more often main line railway's commuter services.[2] Rail in general, and commuter services in particular to places like Pinner, at the core of Metro-land, and Takarazuka, along the Hankyu railway, were relatively expensive as there was no feasible alternative. Those commuter trains catered mostly for the relatively rich who could afford to live in the outer suburbs, closer to nature. Many of the rich and the middle classes lived in apartments in the core of the city, especially in continental Europe. English speaking countries favoured less dense living arrangements.

In any case, the poor were either packed in older, dirty and badly maintained areas very close to the city centre, in shantytowns, slums or in newer low rent developments in the city core fringe or close to industrial areas.

Both the public opinion and most of the *intelligentsia* of the time saw the modern, industrial, mechanised city as a terrible and dehumanising place to live. For instance, the nineteenth century and early twentieth century German sociologists (Marx, Tönnies, Durkheim, Weber, Simmel...)[3] see the contemporary city so problematic as to be doomed. Even if Tönnies differentiates between *Gemeinschaft* (community) and *Gesellschaft* (society) he despised the city of his times and found it alienating. Weber thought cities could be positive, but looked at cities in the past with a romantic eye. This is a common idealisation of the past. We are left with mostly good examples of past architecture, not the bad ones. Pyramids and cathedrals were made to last, and they did. Many of the best nineteenth century buildings still stand, proudly, in most European city centres. There were many low quality, ugly buildings in the olden times that were simply demolished. So, both Max Weber and us know just a sanitised version of the past. For instance, an attentive look at Piranesi's *vedute* of Rome reveals Ancient, Renaissance and Baroque monumental buildings standing amidst muddy roads. That was how things were at the time pretty much everywhere Europe as universal paving did not become a need up until the automobile era and mud mixed with animal excrement was a fact of city life. One could even say that the Trajan column is a contemporary monument as it stands in the middle of pavement, is eroded by internal combustion engine fumes and photographed by digital cameras. In an interesting twist, Hans Christian Andersen wrote a nice tale about the perils or idealising the past in *The galoshes of fortune*.[4]

---

[1] Or elevated as in Berlin and many American cities.

[2] In some cases, those services were being electrified, i.e. the lines that became the Southern Railway and provide commuter service to Surrey, Kent and Sussex.

[3] See any urban sociology manual, for instance Monti et al. (2014).

[4] "It a literary fairy tale about a set of time-travelling boots. /.../ The fairy tale concludes with the clerk's neighbour, a theological student, asking for the galoshes. The clerk gives them to him. As the student walks away he wishes he could travel to Switzerland and Italy, whereupon he is on top of the Mont Blanc. In the freezing weather he wishes he was on the other side of the Alps, where he ends up in Italy, near lake Trasimeno. There he enjoys the beautiful landscapes, but he is confronted with the local people's hunger and poverty. He concludes that it would be better off if his body could rest, while the spirit flies on without it. The galoshes grant his wish and he is now peacefully

Fig. 8.1 Hubby ends up crashing with a streetcar in Hot Water (1924). Pathé Exchange Inc. 1924 publicity photo. Public domain archival image retrieved from http://postalesporinternet.blogspot. com/2011/03/news-56538-56547.html)

It is in this context that a new transportation technology, the motor car, comes to revolutionise the way people move around and will end up determining how cities work. For almost three decades cars were just toys for the rich. They only become to be used at a large scale after the end of the 1914–1918 Great War.[5] However, once it started, mass motorisation advanced at a very fast pace. That was the case in the United States. By the 1920s cars like the Ford Model T were mass produced and affordable to the vast, growing American middle classes. All of a sudden, the streets that just a ten years ago were congested by pedestrian, carts and most of all trams were now full of cars. In *Hot Water* (see Fig. 8.1), a famous silent film from 1924, Harold Lloyd's character (Hubby) marries the girl at the beginning of the movie to

---

dead. Andersen concludes with a quote by Solon: "Call no man happy until he rests in his grave." Dame Care then tells the other fairy that her predictions have indeed all came true. Though she does grant the student a favour. She takes off his galoshes and takes them back with her, causing him to be brought back to life." Andersen (1949).

[5]Motorised road transport played an increasingly important role during the war. For instance, the entire fleet of Paris taxis was mobilised in 1916 to bring soldiers to the frontline at Verdun, Bernede (2006), although the bulk of supplied were brought by a narrow-gauge railway parallel to the Voie Sacree road. By the end of the war, thousands of surplus military trucks came to compete and caused the demise of many local railway lines.

Fig. 8.2 Manhattan mayhem c. 1920. Trams carry much more people than private cars in this open domain picture. They did not last much longer. Public domain archival image retrieved from https://skyrisecities.com/news/2016/08/once-upon-tram-century-trolley-dodging-manhattan-brooklyn-and-queens

then move to a nice house in a tramway suburb. Hubby takes a streetcar (possibly a Pacific Electric Red Car) to his job downtown. However, he is often compelled to carry back home groceries by his wife. Once, after a long shopping visit to the market, he wins the first prize in the raffle: a huge live turkey. He needs to bring back home the whole loot in a very crowded rush-hour tram. The turkey causes mayhem by pecking the backside of a matron. Hubby gets kicked out of the tram and has to walk back home. Utterly frustrated, he buys a Chevrolet Superior Sedan, called a Butterfly Six in the movie. The car turns out to be a lemon for comedic purposes, thus the name change.[6]

Rapid motorisation in America was indeed the spirit of the times and indeed, according to Paul Mees,[7] even if trains and trams moved four times more people than cars in the US in 1920, by 1925 so many Hubbies had bought a car that mileage travelled by car doubled the mileage travelled by rail.

The huge increase in car ownership in 1920s America created a new need for parking space for the growing fleet of cars (see Fig. 8.2). Hubby and others lived now in houses in tram suburbs that already came, or were somehow easy to retrofit, with a garage and a driveway. There were far more problems at destination. The dense

---

[6]See Bengtson (2011) for a summary and commentary of the film.
[7]Mees (2009).

American downtowns with shopping streets and skyscrapers full of offices were not ready for the sudden tide of cars to come in. Parking was a problem in Manhattan more than anywhere else (Fig. 8.3). The New Yorker magazine often satirised the situation in surreal cartoons proposing ever more outlandish ideas to deal with the parking situation such us stackable wedge-shaped cars. More realistic attempts to confront the downtown parking crisis included the installation of parking meters[8] and the construction of parking houses.[9] The congestion and the need of parking space, together with poor management and the new bus technology put such pressure on tram and interurban lines that by the 1960s they have almost disappeared from most American and European cities. However, American architects and urban planners found a better way of adapting downtown to the motorcar. That is, to get rid of it altogether.

From the mid twentieth century onwards newly built cities and city extensions have pretty much everywhere been designed, first and foremost, with the needs of motorists in mind. Old, consolidated city centres were adapted as much as possible all around the world and, in some extreme cases to be found particularly in the US, have almost disappeared or lost their original function as city cores. Notwithstanding problems of pollution, congestion, land use and amenity the car-based city has brought huge gains in terms of economic growth. Several generations in developed countries have only lived in that kind of city so, for them, it is not easy to imagine an alternative.

Tramway suburbs and garden cities were the first to adapt to motorisation. The original idea of the garden city proposed by Ebenezer Howard in his book *To-Morrow: A Peaceful Path to Real Reform* (Howard 1902) includes a fast railway connection to the metropolis plus an efficient electric railway linking the garden cities located around the metropolis with each other. People were supposed to mainly walk or ride bicycles within them. Automobiles made the garden city much easier to implement. Once most people can drive their cars to the city, they can use them for trips inside the garden city or travel to another garden city just using existing roads. So new railways or tram lines don't need to be built anymore, not even bus service may even be necessary at all. Also, garden cities can now grow much bigger. The rail-based commuter town evolves into suburbia in earnest. The prime planning example of this early suburb is to be found in Radburn[10] (Fig. 8.4) where the back of the house has road access and the front of the houses face each other over a common yard with pedestrian paths. Vehicular and pedestrian separation is thus achieved. Another feature of the car-based suburb is a very strict zoning regime. That is, residential areas are to be kept away from retail and commercial areas in general. Zoning also stipulates floorspace ratios to keep population density low. That makes some sense as cars and car use were recognised as a source of negative externalities. Cars are noisy, pestilent and dangerous, so it was thought sensible to keep them away from pedestrians. Most people would be driving to the shops, which would attract traffic and thus become undesirable residential areas, so it makes sense to put them away

---

[8] Shoup (2017).

[9] An interesting example of an early parking house can be found in Venice, near Piazzale Roma.

[10] See, for instance, Birch (1980).

Hotel for Autos

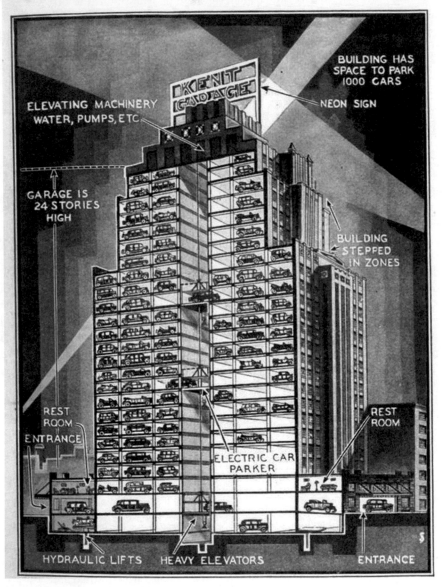

Fig. 8.3 The New York Kent Garage Hotel for Autos is not a cartoon from the New Yorker, but a reality back in the 1930s. It was converted to a warehouse in 1943 and it is now an apartment (condominium) building. Modern Mechanix magazine, May 1929. Public domain archival image retrieved from https://www.citylab.com/design/2015/09/the-marvellous-history-of-new-yorks-hotel-for-autos/405832/

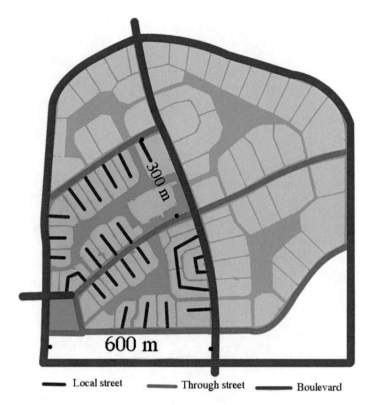

Local street ━━━ Through street ━━━ Boulevard

Fig. 8.4 Road hierarchy in Radburn. Local streets don't connect to each other but end in *cul-de-sacs*. Sourced from Wikipedia

from houses, at least the better-quality ones.[11] Also, population density needs to be kept low as higher density comes with more driving and therefore more undesirable traffic and possibly congestion.

Americans, particularly the white middle-classes, started to move in droves to newly developed subdivisions after the end of Second World War. Those new neighbourhoods were composed of tract housing, also known as cookie cutter houses, and for the most part did not include a separate network of pedestrian paths. Sidewalks were still in place to allow residents to reach the road to be picked up by a car. Most houses included a double garage, and as many as possible were built around a *cul-de-sac*, the most expensive and desirable location within the subdivision (Fig. 8.5).

Tract housing was relatively affordable, definitely within reach of the average middle-class family but critically out of reach of the poor who stayed in now increasingly dilapidated areas around downtown or moved into tramway suburbs, which

---

[11]The original suburb of Radburn achieved some success in terms of use of the pedestrian infrastructure, by the early 70s almost have of the residents were actually walking to the shops. That's very unusual for residents of American subdivisions.

Fig. 8.5 Americans aspired to houses like the one Mr and Mrs Blandings'. A promotional still from the 1948 Hollywood comedy 'Mr. Blandings Builds His Dream House' starring Cary Grant and Myrna Loy. Sourced from Wikipedia

then became dilapidated. Many of those poor were blacks. Suburbanisation encouraged, sometimes explicitly, and resulted on racial segregation. Downtown became poor, ridden with crime and populated by blacks and other minorities.[12]

The *cul-de-sac* (see Figs. 8.6. and 8.7) is the end of the most local road, at the bottom of the road hierarchy in which local roads feed collector roads, collectors feed arterials and arterials feed motorways. This hierarchy, together with the preference for maximising the number of coveted *cul-de-sac*, results on a trunk-branch-twig road structure easily seen in maps. This road structure minimises traffic in the twigs, but it also makes life very hard for pedestrians. A walking trip necessitates of many twists and turns.

---

[12]The word *urban* is often used as a synonym of *black* in modern American English, for instance in *urban contemporary* as a radio format or the use of *urban culture* as an euphemism for African American culture or more precisely Black American culture.

Fig. 8.6 Tract housing with *cul-de-sacs*, no shops or cafes in sight and a double carriageway with no pedestrian crossing. Sourced from Wikimedia Commons

Fig. 8.7 A suburban cul-de-sac in California, this photo shows the ideal "leafy" suburb. Sourced from Wikipedia

With time, the separation of residential and commercial uses was complete. Most shopping moved inside shopping malls surrounded by parking or in strip malls with outdoors parking located along arterial roads. Neither can be accessed on foot. Many jobs, especially the best paid, are still downtown which has been accessible by freeway, but there are also newly developed business parks near freeway junctions. That is the so-called Edge City, see Garreau (1992).

Planning practice and the legal framework also evolved to accommodate car ownership. The strict American zoning regime is known as the Euclid system. The Village of Euclid, Ohio v. Ambler Realty Co.[13] is a 1926 is a United States Supreme Court landmark case. That is, a court decision that establishes a significant new legal principle or concept or otherwise substantially change the interpretation of the existing law. Ambler Realty owned land in the village of Euclid, located in the outskirts of Cleveland. Ambler aimed to use the land for industrial use. The village of Euclid enacted a zoning ordinance which made impossible for Ambler to use their land for industrial uses. The ordinance included provisions for the use of land, heights and area. Ambler sued Euclid claiming that their zoning ordinance was actually aimed at preventing "the colored or certain foreign races from invading a residential section". A lower court decided that the ordinance was indeed unconstitutional in the basis of racial segregation. Euclid took the case to the Supreme Court, which sided with them. The Court reasoned that the zoning powers were not an unreasonable extension of the village police powers. According to the Supreme Court, Ambler failed to prove that the zoning ordinance has a discriminatory character and had no rational basis. Zoning ordinances were thus deemed constitutional in the United States. It is worth noting that the Euclid zoning ordinance was not the first attempt to enact such law. However, the received opinion was that zoning was an unreasonable intrusion into private property rights for a government to restrict how an owner might use property. That all changed for good with Euclid. From then on zoning became to be understood as a form of nuisance control and therefore a reasonable police measure. Further than that, the Supreme Court also argued that that there was valid government interest in maintaining the character of a neighbourhood and in regulating where certain land uses should occur. This decision had a huge impact and soon most cities and towns in the US enacted zoning ordinances. It is interesting to realise that in its origins zoning did not have much to do with facilitating car use. It was indeed more about preserving the character or a neighbourhood and most likely zoning ordinances were enacted as a sophisticated, constitutional way of achieving racial segregation. Car dominance was, once again, an unintended consequence that, with time, become a staple of suburban American culture.

It is possible to understand Euclid zoning as a way to deal with negative externalities. Indeed, right or wrongly, Euclid residents viewed racial integration as a threat to the character of their village. Lower density prevents the construction of large tenements necessary to house factory workers, many of them being blacks and many others recent immigrants to the US. This rule can be defended as protecting the village's bucolic character, and thus it withstood the court challenge. Of course,

---

[13] A more detailed account of the story can be found in Fluck (1986).

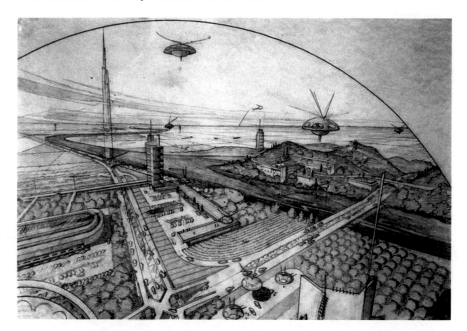

Fig. 8.8 Broadacre, from the public domain, updated for the '50s by Wright with flying saucers galore. Public domain archival image retrieved from http://pc.blogspot.com/2005/11/frank-lloyd-wright-broadacre-city.html

other externalities are to be considered, for instance the development of industrial estates would bring pollution and noise.

We believe that certain zoning rules, in particular those mandating the separation of polluting and noisy heavy industry, make good sense. Euclid zoning, however, goes too far by also mandating the separation of commercial and retail zones from residential zones. Further than that, we also disagree with the low density and exclusion resulting from most American zoning practice. Nobel laureate Joseph Stiglitz agrees, Stiglitz (2019), but this is far from the consensus among American urban economists. The mainstream is closer to the cheerleading views on zoning expressed by William Fischel in *Zoning rules!*, Fischel (2015).

Only a few years after Euclid zoning was finally accepted by the Supreme Court the architect Frank Lloyd Wright came up with his Broadacre City concept, Wright (1932), see Fig. 8.8. As many other architects before and after his time, Wright, a talented designer of buildings, saw himself as capable of designing cities just as well. His landmark idea was to give each American family an acre from the federal land reserves to build a homestead. Zoning is implicit and obvious when looking at the Broadacre model built by Wright's students, but Wright was explicit in envisioning the future of America dominated by car. Walking is restricted to the acre each family owns. Everything else necessitates a car. Broadacre was probably somehow influential at the time it first came as a model and a series of drawings and written explanations. The model was indeed exhibited at the Rockefeller Centre in

Fig. 8.9 Farmland in Indiana follows Jefferson's 1785 Land Ordinance grid. Aerial photo sourced from Wikipedia

1935 and in Pittsburgh only a few months after in an exposition sponsored by the Federal Housing Administration. Even if Broadacre can be seen as a slightly utopic plan it did epitomised the spirit of the times in terms of how human settlements should evolve. Wright went over great lengths to see his plans made into law to a great detail. Other than refining and publishing his ideas until he died in the 1950s he wrote countless letters to politicians, including the US president. He also sought the support of public figures going as far as including the signature of Albert Einstein, against his will, in one public petition sent to president Roosevelt, see Cameron (2014).

Broadacre can be also seen as an update of the 1785 Land Ordinance which divided the lands west of the Ohio river in a grid of townships 6 square miles in size. This ordinance was in line with the agrarian ideas of its main proponent, not other than Thomas Jefferson. Indeed, the Jeffersonian ideas of agrarian democracy and natural rights, see for instance Griswold (1946), are a crystallisation of American cultural views preeminent to this day and very much visible in the strong support for suburbanisation and local provision of public goods. The Land Ordinance grid is a reality, easily seen this day in maps and aerials photographs of the Middle West (Fig. 8.9). In practice, federal laws enacted during the 1930s New Deal and the resulting institutions such as the Federal Housing Administration (FHA) had a much deeper impact than Broadacre and arguably more than the Land Ordinance.

Together with the Federal National Mortgage Association (FNMA or Fannie Mae) the FHA was founded to help jump start the American housing market which collapsed as a result of the Great Depression. The FHA set standards for construction and underwriting and insures loans made by banks and other private lenders. Fannie Mae's original purpose was to provide local banks with federal money to fund mortgages. Both the FHA and Fannie Mae were instrumental to channel public and private funds to build suburbia. Until the late 1960s the FHA actively, if implicitly, discriminated against blacks and other poor minorities. For instance, between 1945 and 1959 African Americans received less than 2% of all federally insured home loans. As a result, the value of inner-city properties decreased and pushed even more whites to the suburbs. This process, known as redlining,[14] was also helped by private sector real estate agents who would cheaply sell a property to a black family in a predominantly white inner city neighbourhood (often a tramway suburb) causing panic so the white families would sell at a low price to buy (maybe through the same agent) a property in the suburbs.

The rise of the predominantly white, car-based American suburb had a deep impact in the profitability of public transport enterprises. Most commuter railway services, interurbans, tramway lines (streetcars) and even the New York subway were run as for-profit private enterprises in the early-mid twentieth century. Some regulation was in put in place by city and state governments mostly to stop the rapacious transport monopolies from gauging the public. However, the writing was in the wall: the profitability of the public transport business was severely hit by mass motorisation in the 1920s. Many private transit companies simply went bankrupt as a result of the Depression and necessitated government assistance to keep operating. As it often happens with monopolies, public transport companies were not agile enough to face competition. Some obtained cash injections from city authorities. Most were took over and kept operating as public enterprises.

By the late 1930s the both bus and coach technology were mature enough to be seen as a replacement of streetcars and interurbans. Buses were seen as sleek, modern and fast when compared to turn-of-the-century streetcars running in badly kept track. It is worth noting that at the time there were already newer streetcars[15] comparable in comfort and speed to buses and, critically, offering significantly higher capacity than buses especially when run in coupled pairs. What could have been a significant rebirth of urban rail services came to an end in many US cities as a result of the so-called General Motors (GM) streetcar conspiracy. By the late 30s a group of companies led by GM and including some of the most important oil and tire (Firestone) producers put together National City Lines (NCL). Subsequently and during more than a decade NCL and its subsidiaries bought struggling streetcar companies across the US and, in most cases, converted them to bus operation. Of course, buses were supplied by GM

---

[14]See Rothstein (2017) for an excellent summary of the history of redlining to this day.

[15]Such as the streamlined Presidents' Conference Committee (PCC) cars and the Brilliners manufactured by J. G. in Philadelphia. Earlier in the century Brill exported interurban technology to Europe and Japan. PCCs were in relatively widespread service with the best streetcar companies by the mid 1930s. Some are still in revenue service nowadays.

and used petrol and components manufactures by the other owners of NCL. Most of the companies involved were convicted in 1949 of a conspiracy to monopolise interstate commerce in the sale of buses, fuel and supplies to NCL and its subsidiaries. It is unclear whether GM had the intention of monopolise and downsize the transit industry to further encourage the development of the car-based suburbia, as some conspiracy theorists suggest.[16] It may well be true, but as we have argued above many other powerful forces were pushing hard for suburbanisation at the time. Another of those forces, which we have not mentioned so far, was the extensive federal funding devoted to the construction of the Interstate Highway System from 1956 onwards. Note that this nationwide network of high-speed freeways helped to channel traffic from the suburbs into the central business districts across America. Cities and States developed, with substantial help from the federal government, their own freeway systems to form a car nation.

American cities further adapted to the reality of car dependency by refining their approaches to planning in general and zoning in particular. Chiefly among those stands the ubiquitous is the "parking minimum requirement", see Shoup (2017) for anything about parking in the US. By some estimates, there are eight parking spaces for each car in the US. The lowest estimation puts the proportion at three parking spaces per car. Neither includes home parking. This oversupply of parking, generally free of charge, is once again a result of policy rather than *laissez-faire*. The majority of American cities started to impose a minimum parking requirement to new developments, commercial or residential, back in the 1930s. At the time street parking was scarce so municipal governments decided to regulate it with parking meters. That was not an acceptable solution for new developments as they needed more parking that could be accommodated in the street in any case. So, municipal ordnances started to mandate a number of off-street parking spaces for the use of customers and visitors. Nowadays, US planners use the Parking Generation Rates to calculate how much parking a particular development needs to have. Parking generation rates are calculated relative to the peak parking observed in the area and taking into account the floor area. For instance, a typical restaurant, a stand-alone single-storey structure given the zoning rules separating commercial from residential uses, often needs as much as three or four times more space for off-street customer parking than for the building itself. That has a huge impact land use, and the way the city looks. Walking to look for a restaurant becomes frustrating as the sidewalks are continually interrupted by entrances to the parking lots. Customers must then walk across the car parks to see the restaurant menu and walk back to the sidewalk to continue their search. That makes a city unpleasant for pedestrians and creates just the opposite of an activated street. Pedestrians would eventually next to disappear in this environment. Business will hold substantial market power: if you don't like the price or what is in the menu you need to drive somewhere else. Offers will be monotonous: people go to MacDonald's or other franchised restaurant because they

---

[16]The most popular of those being the 1998 film *Who Framed Roger Rabbit*. See also Slater (1997) for a more serious account of the GM streetcar conspiracy.

know what to expect so they save the search cost,[17] see Schlosser (2012). Note that making mistakes in choosing where to buy a service in a car-based city are costlier as search is done by automobile rather than walking. Moreover, there is an effect in the size and density of the city and a visual effect, business need huge billboards so they can be seen by motorists travelling much faster than pedestrians.[18] All the fine grain of the old city is lost. The editors of the London magazine Architectural Review saw what was coming already in the 1950s: *The USA might conceivably go down in history as one of the greatest might-have-beens of all time. This could be an unsuccess story of a gigantic sort and, in that sense, pleasingly American, but in every other sense a tragedy, a major tragedy for civilization, unthinkable were it not that the symptoms are there, can be scrutinized, diagnosed—just how is suggested in this issue of the Review.*[19]

The existence of mandatory parking affects the amount of land available for other uses and therefore increases its cost. This problem is especially relevant in cities with high housing costs precisely like the most successful ones on both the East and West coasts. The cost of mandatory parking minimums is not paid by customers directly but by developers who would then charge them indirectly to society as a whole. In words of Donald Shoup:

> Off-street parking requirements collectivize the cost of parking, because they allow everyone to park free at everyone else's expense. American drivers park free at the end of 99 percent of all their automobile trips. If the cost of parking is hidden in the prices of other goods and services, no one can pay less for parking by using less of it. Off-street parking requirements thus change the way we build our cities, the way we travel, and how much energy we consume. All the required parking spaces spread the city out, and the greater travel distances make driving almost a necessity. Shoup (2017)

A parking space is not a public good. Although it can be used by different cars at different times it is still a rival good. Street parking is also excludable, and excludability can be enforced by parking meters and police tow trucks. However, in many jurisdictions street parking is free. That is excludable, but not excluded. The idea is that, once commercial developments are obliged to provide free parking, street parking can be also provided for free to generate a positive externality for motorists. On the other hand we wonder how such a collectivist way to organise parking, one that allows people to leave their private property for free on land they don't own, can be the rule is the free-market paradise that America purports to be.

Exclusion is still very much at the core of the American approach to urbanisation. It is no longer brash racial exclusion, for instance affluent and successful Asian Americans have a good chance of being accepted in richer neighbourhoods, but essentially exclusion of the poorer. The root cause for that is the way public goods are provided in the US. That is, for the most part, they are provided and enjoyed locally. Even schools are funded by a tax on property raised in the school district

---

[17]Or in other more American words, McDonald's is *convenient*, a staple.

[18]In other words, search costs in a car-based city are higher. Sellers will try to minimise those by making themselves more visible.

[19]The text is probably better known from the book *The Australian Ugliness*, Boyd (2010).

rather than general taxation at the state or federal level. The same, of course, is true for the fire brigade, parks, sport fields, street maintenance, water supply and critically the police force. Therefore, an affluent town can provide high quality public goods to its residents who in turn have strong incentives to keep the poor away, see Alesina et al. (2001). Opposition to public transport projects stems from the similar causes, chiefly the fear that apartment buildings will be allowed near a railway station and that trains will bring "bums." Zoning is also largely decided locally and it of course reflects and enforces exclusion, in many cases not only of the poorer but of anyone as increasing density, even with rich people, imposes a strain in infrastructure and could affect the provision of local public goods, at least in the short term. These policies of exclusion have also an impact in the provision of safety. In the most extreme cases, gated communities are surrounded by a wall, with entry controlled by armed guards. The same guards, rather than the police, are responsible for security within the compound. More generally, the local police has orders and incentives to stop suspicious individuals, meaning pedestrians especially the ones of the wrong skin colour.

The reality reflects our assertions in the previous paragraphs. If everywhere in the world we see large population increases in and around successful metropolis, that is not the case in the US. Actually, neither population not economic activity as measured in their share of gdp is increasing in the American star cities. GDP per capita, on the other hand, is growing fast in San Francisco, Boston, New York and other successful American metropolises. How is that possible? The answer is that those cities are home to a shrinking share of the overall population.[20] That is, because the housing stock in those cities is not allowed to grow, the price of real estate becomes exorbitant. Only the very rich computer engineers, biochemists and financiers working in the most successful industries are then able to afford living there. Well, those and the homeless. The middle class is expelled, and the services provided by middle income workers become also extremely expensive, as a haircut in San Francisco. Those workers move to the Sun Belt cities, such as Fort Worth or Omaha, where they will get a far lower salary, but they will be able afford a large house in a cookie cutter suburb.

8.1 The Logic of Congestion

So far, we have described a contemporary car-based city that, in America, is rooted and still fosters economic segregation, often correlated with race. It also generates large amounts of pollution and is energy inefficient. Perhaps most importantly we have argued that the car-based city limits the variety of goods and services on offer and, finally, generates a bland, repetitive landscape made to the scale and speed of the automobile. A sensible question at this point is whether such a city can expand and keep generating wealth in the long run. If so, as much as we dislike it, such model of

---

[20]Yglesias (2019).

growth would be here to stay. Would such a city be able to cope with congestion in the long run? Is building more roads and providing more parking a viable long-term strategy?

Many studies show how building more roads does not ease congestion in the long run. There is economic logic to that. Let's consider a four-lane motorway going downtown from, say, the Eastern suburbs. The motorway is utterly congested in the rush hours so it can't cope with more traffic. People have serious doubts about moving to the Eastern suburbs because of the traffic situation. Many residents avoid driving during the rush hour as much as they can. Some even catch a bus to the railway station. Even if buses and trains are rather slow, the public transport commute is about as long as driving for many people. Deliveries and other commercial traffic is scheduled in the off peak to avoid the traffic jam. Developers don't want to invest there. Then two more lanes are added to the motorway. Congestion is busted. Traffic can now flow much better and a 1 h commute is now cut to a speedy 30 min drive. More people choose to drive, even during the rush hour. Children are driven to schools further away. Deliveries can now be scheduled reliably. Public transport patronage plummets and services are cut. Now families are suddenly attracted by the charms of the Eastern suburbs and the ease of the commute. Developers take note. Within a few years' population grows in the East. Families have moved into their new houses in the Far Eastern suburbs. But with the increase of population congestion came. The six-lane motorway is congested in the rush hour. It now takes 1 h and 10 min to get from the Eastern suburbs to the city, and 1 h and 30 min from the Far Eastern suburbs if you are lucky.

The phenomenon described in our example is a pattern found by many studies, see Duranton and Turner (2011). In the same vein, the Lewis-Mogridge position observes that as more roads are build more traffic fills these roads and speed gains cease to exist in a matter of months, see Lewis (1977) and Modridge (1990). Whenever a speed gain is realised some other road of junction becomes more congested so the system as a whole stays pretty much where it was in terms of average speed in and congestion in the rush hour. Otherwise, the motorways offer off-peak trips. The idea that traffic rapidly expands to meet the available road space is often called induced demand, or even the "Iron Law of Congestion". Further than that, the Downs-Thompson paradox states that the average speed in a congested road will match, in equilibrium, the average door-to-door speed of the trips made by public transport. As a consequence, building more roads will not ease congestion as induced demand will fill them soon enough up until their average speed falls down to match public transport door-to-door times.[21]

---

[21] The Downs-Thomson Paradox, originally thought and tested for London, applies to congested cities with extensive, effective public transport.

8.2 How About Smart Cars?

Electric autonomous cars for hire are unlikely to replace car ownership to a meaningful scale. In a car dependent city, pretty much each person over 16 years old has immediate access to a car. Usage peaks in the rush hours when thousands of commuters, often riding solo, drive from home to workplace or vice versa. The idea of rental autonomous cars is that they would be continuously taking passengers to their destinations rather than sitting idle in car parks. The problem is that, with car usage concentrated in the rush hour, at least one rental car per commuter would be necessary. Because usage would be much lower after the peak autonomous rental cars would just sit idle in the same way private petrol driven cars sit idle nowadays. Things could be improved if commuters would agree to carpool rental shared cars, but we are afraid this would not go much farther than car-pooling goes with conventional cars. Another idea would be to introduce on-demand, autonomous mini-buses to get people to railway stations. That may potentially work as an improvement over (autonomous or not) buses on poorly scheduled routes, but if such routes are not poorly scheduled the improvement could be noticeable but lower than expected. Note that on-demand mini-buses work well in some cities, notably Hong Kong, but are absent from most.

Another usual argument in favour of autonomous cars is cost. Those cars would be much cheaper than taxis or shared cars (driven by humans) like Uber thanks to savings in labour cost. Estimates indicate labour cost account for about 40% in the taxi industry. Would a potential 40% discount in taxi fares make people to ditch car ownership? We doubt it. The average commuting time in Los Angeles is about 55 min. If this is split in two commutes it would result in $56 spent on the cheapest UberX fare per day. A 40% discount in this fare results in about $34 per day spent in commuting in an autonomous car. Over a year with 261 working days that amounts to $8874. According to some studies only 26% of car usage is commuting. Even if 50% of car usage were commenting an average car dependent person in Los Angeles would be spending at the very least whopping $17,784 a year in autonomous cars and more likely something above $25,000. A new Honda CR-V, the best-selling car in the US, costs only $24,000 with about $4000 needed to pay for gas, insurance and other costs.[22] If anything, the cost of car ownership has been falling relative to overall income for the last decades. We believe that a far more likely, and disastrous, scenario is one in which people buy autonomous cars. Self-driven cars could be sent back home to park or sent around the city to look for a parking spot. Having cars running twice as much would most likely more than offset any potential environmental gains of electric power (after all most electricity is produced by burning fossil fuels). Having cars parked all over will result in no gains from the current situation either. The only gain of having electric autonomous cars would be in terms of reduced pollution where the car is driven, and a likely modest gain in overall pollution. Congestion, land use and other problems would not be improved at all.

---

[22]Estimates elaborated by the authors and based on prices and other data available on-line.

At least in the short term, it feels like excitement about new technologies rather than true potential is behind the idea of autonomous cars replacing private car ownership. We don't know what the future may bring us. Autonomous cars may become very cheap to run and operate. They will, however, become cheap for both companies and individuals. It may well require some sort of regulation to stop people from buying them.

8.3 Modern American City Thinking and Trends

So far, we have explained how the contemporary car-based American urbanisation works, and how they got there. Several schools of thought aim to change the urban environment in one way or another. We will now provide an overview of those ideas.

8.3.1 The Death and Life of the Great American Cities

Jane Jacobs published the book that serves as a title to this section back in 1962 when the perils of motorisation were not obvious to everyone, for a recent edition see Jacobs (2016). After the end of WW2 Jacobs, a journalist, and her husband, an architect, decided to challenge the trends by not moving to the suburbs and instead settle in Greenwich Village. The Village, located in lower Manhattan, is one of the few areas of New York in which the streets don't follow a grid pattern. The Jacobses very much liked the local atmosphere of the walkable neighbourhood with little shops. Greenwich Village was, however, threatened by the construction of the Lower Manhattan Expressway, an elevated motorway. There were also plans for "urban renewal" and "slum clearing" in the area. Jane Jacobs became very vocal in her opposition to those plans. As a journalist, and being the wife or an architect, she was giving the opportunity to express her views Fortune magazine, where she published "Downtown is for people" in 1958. Jacobs developed her own ideas about urban life and published them in 1961 in the "The death and life of great American cities".

Jacobs emphasises the need to allow the city to adapt to complex human needs by fostering, or simply let happen, diversity and mixed use. For that, she argues, sufficient density is paramount. Connectivity, in the form of small blocks not necessarily forming part of a grid is also important in her eyes. Jacobs also emphasised the need for the city to be safe. She coined the phrase "eyes on the street", which in her own words stand for an "intricate, almost unconscious, network of voluntary controls and standards among the people themselves, and enforced by the people themselves." She also emphasised the importance of the sidewalk and the small public park as places for staying and meet others, for strolling aimlessly, play or window shop. Although one can see a clear tint of parochialism in Jacobs opinion, something akin to "it cannot get better than the Village", she goes much farther than that by providing a well organised, demolishing critique of suburbanisation, the CIAM principles for urbanisation and specially the urban ideas of Le Corbusier. She, in our view correctly, identifies the CIAM/Le Corbusier design as actually guided by aesthetic consideration and an oversimplification of human needs to make room for modernist towers in a park. Although she was more sympathetic to the Garden City ideals, she saw those as a precursor to the soulless suburbs sprouting in the 50 s all over America. Interestingly, she was also critical of grand schemes in general. For instance, she deemed the efforts of the City Beautiful movement unsuccessful. This is one of the reasons Lewis Mumford, otherwise an ally in her opposition to the Lower Manhattan Expressway, accused her of aesthetic philistinism. It is planned public spaces and public buildings what makes cities like London and Paris great, Mumford argued.[23] We would argue that the scale is what makes public spaces or monuments to work in a particular city, and that urban design is best when not tied to specific aesthetic consideration. Let the architect be responsible for how a particular building looks. Coming back to monuments, we believe that easy pedestrian access and allowing for multiplicity of uses more around than in a monument make it work. That's why, for instance St Stephan's cathedrals "works" better than any of the large monumental buildings along the Ringstrasse in Vienna, surrounded by barren, unappealing open space. One walks to the cathedral but uses a tram to go around the Ringstrasse. Things are much worse if access is mainly possible by private car, of course.

In any case, Jacobs started several important themes in planning, urban design and urban economics. Her influence can, by no means, be underestimated.

8.3.2 Paul Mees, Public Transport for Suburbia

The late Paul Mees was an Australian academic who wrote a Ph.D. thesis and several books about public transport and planning in the suburban context that dominates most of the urbanised areas in Australia. His arguments and ideas are particularly interesting as it is easy to think that they extrapolated to other parts of the Anglosphere, the US in particular.

Mees's main argument is that low density suburban areas can and should be served by effective public transport to the extent that it could be the dominant mode. He shows how a country such as Switzerland, with no large, dense urban core in it provides universal access to public transport which in turn is well-patronised, well-liked by most people and by and large not a huge burden for the public coffers.[24] Mees compares the Zurich system to that of Melbourne in the 1990s, where even a simple transfer between train and bus to access suburban Monash university was an ordeal. He also argues that things used to be better in terms of public transport for the suburbs. For instance, in the 1950s Auckland's or Sydney's transportation

---

[23] Mumford (1986).

[24] Notably public expenditure as a percentage of gdp in Switzerland is well below the OECD average and even lower than of Australia.

to the suburbs was dominated by rail, trams and buses rather than private cars. It was the municipal and state governments who, in light of increased car ownership, decided to invest in roads rather than modernising public transport. That is a story similar to what started in the US before the war and culminated in the late 1940s. However, things happened without the huge push factor of racial discrimination in Australia and New Zealand. Perhaps because of the latter factor and also because of a planning system more centralised than in the US, some eminently suburban cities in the English-speaking world retained a far better public transport system. Mees cites as an example the successful and efficient bus system in the relatively small Ottawa and gives also praise the relative success of Toronto when compared to Melbourne. Some recent and not so recent developments in Australia and Canada can be used as an example of what it can be achieved to provide decent public transport to the suburbs. In particular, we are very impressed with large increase in public transport usage in Sydney which is growing faster than its fast-growing population since 2014.[25] Patronage numbers for the 2018–2019 fiscal year at 773.3 million boardings have already surpassed far surpassed the government's optimistic prediction of 680.5 million by 2031. The surface area of greater Sydney is similar to Los Angeles County, but with only about half of the population a much lower population density. Public transport ridership in LA amounts to 1.6 million for the average weekday compared to 2.6 million for Sydney. That is, public transport usage in Sydney is 3.25 times larger in a per-capita basis than in LA. The largest contributor to Sydney's patronage is a 400 km heavy rail network that reaches far into the suburbs. Access to the largest employment area, the Sydney CBD, is 80% by public transport, which accounts for an overall commuter mode share (to all areas) similar to European cities. Sydney still has a much larger private car mode share than them because people rarely have the opportunity to walk to work given the long distance in such a spread out city and the inconvenience of walking in the outer suburbs littered by noisy, polluted arterial roads where a potentially nice 1 km walk to the shops next to the railway station becomes an ordeal.

We believe that Sydney's success when compared to other Australian and especially American cities relies first and foremost how density is managed. Nobody really talks much about transport-oriented development in Sydney, it just happens at a large scale. Large concentrations of high-density living have sprouted around major transportation hubs such as Chatswood, Parramatta or Green Square. The latter is admittedly quite problematic as the size of the development far surpasses the capacity of the local railway station. A virtuous cycle of infrastructure construction (metro, light rail, bus improvements…) together with an increase of frequency which fosters patronage in the peak, and especially off-peak, is happening in Sydney. The city is experiencing a relatively fast and definitely wide-ranging transformation which, of course, does not come free of problems.[26]

---

[25]Other Australian public transport systems are not experiencing such strong growth.

[26]For instance, Sydney's suburban railway is designed to provide one seat trips for commuters. It is, therefore, are tangled system which relies on branching rather than transfers. That, coupled with the surge in demand, causes serious capacity and reliability problems. The new Sydney Metro

The experience of some Canadian cities is also very interesting. Public transport ridership in Toronto, Montreal and Vancouver is dominated by metro networks that cover a relatively small part of the metropolitan area, in a way more similar to the European experience than to Sydney's. The urban centres of those Canadian cities are larger and more densely populated than in Australia. That probably makes their experience harder to imitate for US planners.

8.3.3   Edward Glaeser's Urban Economics Critique

Urban economics, essentially Americans or shaped by the American academia, traditionally regards the city as a good place to work rather than a good place to live. Increasingly both things are getting somehow entangled, and to a modest extent Americans are re-discovering the city lifestyle. It is clear that outside of America walkable cities offer access to a variety of goods and services that are not available in the car-based city. The existence of those goods may be enough to attract human capital and make the city grow. So, there is an economic value to density and walkability which is recognised by some urban economists. There are a couple of related things to take into account: access to high quality education and safety are among the both in very high demand. The access to those public goods has been much restricted in inner city areas in the US. The reasons are again institutional rather than market driven: the way schools are funded (through property taxes) and the fact the city dwellers are poorer and more prone to the type of crime that spooks families out of the inner city. Both are a great challenge for the transformation of city centres. In Harvard's professor Edward Glaeser words:

> Urban economics has traditionally viewed cities as having advantages in production and disadvantages in consumption. We argue that the role of urban density in facilitating consumption is extremely important and understudied. As firms become more mobile, the success of cities hinges more and more on cities' role as centers of consumption. Empirically, we find that high amenity cities have grown faster than low amenity cities. Urban rents have gone up faster than urban wages, suggesting that the demand for living in cities has risen for reasons beyond rising wages. The rise of reverse commuting suggests the same consumer city phenomena. [...] restaurants, theaters and an attractive mix of social partners are hard to transport and are therefore local goods. Cities with more restaurants and live performance theaters per capita have grown more quickly over the past 20 years both in the U.S. and in France. (Glaeser et al. 2001)

The same piece, titled "Consumer city", argues there are three other particularly critical urban amenities other than a variety of services and consumer goods. Those are: aesthetics and the physical setting (including architectural beauty), public good

---

system should solve some of those problems, but it will take years to complete. There is also rampant NYMBYism which, given the centralised planning, has a lower change of success than in America. However, NYMBYs are still able to lobby their local members, get media exposure, and occasionally achieve results.

services and speed. Note that most of the urban amenities mentioned cannot be obtained in the market.

Alas, most of the urban economics field has a far more conservative approach, littered with mathematical solutions to the so-called "shopping center problem", using theorems and econometrics to provide justification for privately funded police forces and so on.[27]

8.3.4 Richard Florida

The Floridan (by Richard Florida) approach is to wash up the inner city so it remains grungy and bohemian, but it becomes attractive to the "creative" class, see Florida (2005). The idea is that the creative class is endowed, or comes with, the human capital[28] and human capital likes rock bands, lofts, artsy cafes and so on. The Floridan approach has a lot of pop appeal and creative hubs are sprouting everywhere. Glaeser's critique to Florida's ideas[29] hinges in the fact that not all human capital has bohemian tastes. Their views can be easily reconciled by adding a dynamic approach. That is, young childless creative people in the US would indeed flock to the bohemian inner-city areas where they can find a music scene, hip restaurants and clubs and a tolerant environment. Eventually they will form families and most of the bohemians goods will become out of reach. Now, safety, good schools and a big house with a double garage and a backyard become paramount and the former bohemians will move out to the suburbs. Occasionally, when the kids grow up a bit, they will be able to travel, by car, to a parking area next to the bohemian centre and go for dinner to some South Indian place. The bottom line is that bohemian and family goods are more incompatible in America than in other places, because of historical reasons laws and regulations.

8.3.5 The New Left, the Gentrification and Other American Planning Buzzwords

Otherwise, the big lefty criticism to Florida are the accusations of generating gentrification. That is, the idea that fixing public infrastructure in the centre of the city in order to attract bohemians and hipsters ends up in an increase of rent and housing

---

[27]Tyler Cowen's blog is full of such claims and links to the relevant articles.

[28]We find the definition given by Wikipedia good enough: Human capital is the stock of habits, knowledge, social and personality attributes (including creativity) embodied in the ability to perform labour so as to produce economic value.

[29]Glaeser actually seemed to like Florida's book, considering it a good summary of urban economics themes and ideas (Glaeser 2005). However, both he points out there is little evidence to support Florida's claims about the Gay Index and the Bohemian Index. They both become irrelevant after controlling for human capital, see Glaeser and Saiz (2003).

values and therefore pushes the poor out of the area. The poor has to live somewhere, and according to Gleaser, the American poor live in or next to the city centre because this is the only place with some, albeit skeletal by first world standards, provision of public transport. That is supported by Glaeser's regression analysis and is easy to see in places like San Diego if one adventures to use the public transport. Thus, the poor will be much worse off by being forced out to some grotty suburb far away from public transportation. Otherwise, gentrification would not be that bad. And, indeed, measures could be taken to improve the situation. For instance, public or subsidised housing could exist in every street. That seems better than packing, or allowing to live, the poor all together in some undesirable area. Also, universal access to public transport would also enhance the situation. That's one of the reasons Glaeser and Stiglitz support public transport development, to an extent, the other reason being the huge environmental cost of suburbia.

In essence, brutal gentrification is a result of American planning practice. We have argued before that the residents of affluent areas control their zoning to keep the poor away. By the same reason, impoverished city municipal government, usually governing the old and impoverished city centre have incentives to attract richer residents. After all, they will pay taxes that are the main source of income to improve the quality of schools, parks, public transport etc. It is easy, even for a well-intentioned City Hall, to enact policies attracting rich residents to an extent that the poor have to move away. Rather than marching and putting the culprit on neo-liberal market forces, a widespread liberalisation of zoning, allowing for higher densities everywhere and curtailing policies of exclusion, would help to contain gentrification by using the same market forces. In needs to be clear, however, that changing the law and letting the market to the work will not solve the problem. Universal provision of sufficient quality public goods, everywhere and for everyone is also a condition *sine qua non*. This is how things work out of the US, not even that far, just in Canada. The same can be said about Europe, Australia and Japan. The local provision of public goods to the exclusion of others is at the core of American exceptionalism.[30] Not that many people have noticed so far.

A similar logic applies to many of the criticism and remedies to urban ills observed in the English-speaking countries and beyond. American planners have come with a plethora of ad hoc remedies for a regime that is essentially flawed, and they have exported their vocabulary. We have Transit Oriented Development (TODs), Inclusionary Zoning, Smart Growth, Urban Growth Boundaries etc. TODs exist outside of America in a much larger scale, only they are not often call that as they are just business as usual. Inclusionary Zoning does not need to be contemplated if standard zoning is not as excluding by default as it is in America and so on.

---

[30]Other than in public good provision, there is a stark difference in income redistribution between the US and Europe. In general, European countries are much more generous to the poor. There, transfers to low income recipients come mostly from progressive taxation rather charitable donations like in the US. See Alesina et al. (2001) for a detailed explanation.

8.3.6 Light Rail Versus the Kochs and the Great American Public Transport Melancholy

It is obviously not all silly lefty criticisms to the status quo in America. The status quo itself has a large majority and powerful interests behind it. An example is the successful battle of "Americans for prosperity", a group funded and supported by Koch Industries against the Nashville light rail. In truth, modern light rail in the US has a moderate success[31] (and that's being generous) because zoning and other planning practice does not allow for substantial development around stations. One could argue that, if the transport infrastructure is in place, the development can come after. Current statistics are, however, very discouraging. All in all, public transport ridership is in sharp decline in the US. This decline is generally attributed to the success of ride hailing services, such as Uber and Lyft,[32] and used as an excuse to cut services. Most of the decline happens in the bus services patronised by the poor, but even the New York Subway has been recently going through a crisis caused by low maintenance standards which in turn impacts ridership. Note that the strong increase in ridership in Sydney explained before is happening in parallel to a boom in ride hailing services, which have now surpassed taxis. We believe that the core cause for ridership decline in the US is the low quality of service usually designed for the very poor. Once an acceptable alternative attractive to the poor, who can share a ride with each other or the urbanite Floridan hipster frustrated by the horrors of American transit, appears they flock to it in sufficient numbers to provoke a decline in public transport use.

References

Alesina, A., Glaeser, E., & Sacerdote, B. (2001). *Why doesn't the US have a European-style welfare system?* No. w8524. National Bureau of Economic Research.

Andersen, H. C. (1949). *The complete Andersen: All of the 168 Stories.* New York: Heritage Press.

Bengtson, J. (2011). *Silent visions: Discovering early Hollywood and New York through the films of Harold Lloyd.* Santa Monica Press.

Birch, E. L. (1980). *Radburn and the American planning movement the persistence of an idea* (pp. 424–439).

Boyd, R. (2010). *The Australian Ugliness.* Melbourne: Text Publishing.

Cameron, M. (2014). *Albert Einstein, Frank Lloyd Wright, Le Corbusier, and the future of the American City.* The Institute Letter, Spring.

Duranton, G., & Turner, M. A. (2011). The fundamental law of road congestion: Evidence from US cities. *American Economic Review, 101*(6), 2616–2652.

---

[31] In terms of ridership per km modern American light rail is on par with what is simply achieved with buses in Europe or Japan. Trams in some central European cities such as Budapest of Prague move an order of magnitude (that's 10 times more) passengers than American systems. Still, light rail in the American urban left's darling as it attracts some middle-class patronage and oppose to buses, only used by the numerous literally unwashed American underclass.

[32] Graehler et al. (2019).

Fischel, W. A. (2015). *Zoning rules! The economics of land use regulation*. Lincoln Institute of Land Policy.

Florida, R. (2005). *Cities and the creative class*. Abingdon: Routledge.

Fluck, T. A. (1986). Euclid v. Ambler: A retrospective. *Journal of the American Planning Association, 52*(3), 326–337.

Garreau, J. (1992). *Edge city: Life on the new frontier*. Anchor.

Glaeser, E. (2005). Edward L. Review of Richard Florida's the rise of the creative class. *Regional Science and Urban Economics, 35*(5), 593–596.

Glaeser, E. L., Kolko, J., & Saiz, A. (2001). Consumer city. *Journal of Economic Geography, 1*(1), 27–50.

Glaeser, E. L., & Saiz, A. (2003). *The rise of the skilled city*. No. w10191. National Bureau of Economic Research.

Graehler, M., Mucci, A., & Erhardt, G. D. (2019). Understanding the recent transit ridership decline in major US Cities: Service cuts or emerging modes? In *Transportation Research Board 98th Annual Meeting, Washington, DC,* January, 2019.

Griswold, A. W. (1946). The Agrarian democracy of Thomas Jefferson. *American Political Science Review, 40*(4), 657–681.

Howard, E. (1902). *Garden cities of to-morrow: A peaceful path to real reform*. Swan Sonnenschein.

Jacobs, J. (2016). *The death and life of great American cities*. Vintage.

Lewis, D. (1977). Estimating the influence of public policy on road traffic levels in Greater London. *Journal of Transport Economics and Policy*, 155–168.

Mees, P. (2009). *Transport for suburbia: Beyond the automobile age*. Abingdon: Routledge.

Mogridge, M. J. H. (1990). *Travel in towns: Jam yesterday, Jam today and Jam tomorrow?* Berlin: Springer.

Monti, D. J., Borer, M. I., & Macgregor, L. C. (2014). *Urban people and places: The sociology of cities, suburbs, and towns*. Thousand Oaks: Sage Publications.

Mumford, L. (1986). Home remedies for the urban cancer. In D. Miller (Ed.), *The Lewis Mumford Reader*. Pantheon.

Rothstein, R. (2017). *The color of law: A forgotten history of how our government segregated America*. New York City: Liveright Publishing.

Schlosser, E. (2012). *Fast food nation: The dark side of the all-American meal*. Boston: Houghton Mifflin Harcourt.

Shoup, D. (2017). *The high cost of free parking: Updated edition*. Abingdon: Routledge.

Slater, C. (1997). General motors and the demise of streetcars. *Transportation Quarterly, 51,* 45–66.

Stiglitz, J. (2019). *People, power, and profits: Progressive capitalism for an age of discontent*. UK: Penguin.

Wright, F. L. (1932). *The disappearing city*. WF Payson.

Yglesias, M. (2019). *The American economy isn't actually becoming more concentrated*. Vox, 2017. Retrieved in May, 2019.

Chapter 9
The Japanese Experience: The Rise of the Minimal Car Use Megalopolis

Abstract We discuss the Japanese approach to planning in contrast to the American approach. We provide a historical explanation of the evolution of Japanese cities and their planning from the Tokugawa Era to the present day. Although very large, cities in Japan are not that dense. Like in America, the majority of the population live in single-family houses located in the suburbs. However, Japanese planning allows for mixed use by default and forbids on street parking nation-wide. As a result, railways are the main mode of transportation for longer distances within the city, while most errands can be solved by walking or cycling to either a corner shop or the larger commercial and service areas surrounding railway stations. Although with faults and problems of its own, we argue that the more *laissez-faire*, simple approach to planning in Japan is an example to follow that would result in a more efficient, less segregated and sustainable city form.

Keywords Mixed zoning · Japanese planning · Parking limitation · Pedestrian friendliness · Kanto area

The sharpest contrast in terms of planning, public transport usage and motorisation relative to the American experience is to be found in Japan. We will describe the differences, and similarities, between cities in both countries, and explain the cause of those differences.

Considered as a whole, Japan is densely populated. It more than doubles Great Britain in population but its landmass is only about 80% larger. Most notably, population is concentrated in cities along the coast, primarily the Pacific Coast, as inland areas are mountainous and poor. With more than 42 million inhabitants, the Kanto region around Tokyo contains the largest urban area in the world by population.[1] There are other very large megapolises centred in Osaka (Keihanshin) and Nagoya

---

[1] It is actually quite hard to provide an accurate population figure for Greater Tokyo. The Kanto region is a census statistical area comprised of seven prefectures: Tokyo, Kanagawa, Chiba, Saitama, Ibaraki, Tochigi and Gunma. There are, however, substantial rural areas in the Kanto region. The population of the Tokyo prefecture, ruled by the Tokyo metropolitan government, is about 13.5 million. That leaves very large and highly populated urban areas in Kanagawa prefecture, Yokohama,

P. Guillen and U. Komac, *City Form, Economics and Culture*, SpringerBriefs in Architectural Design and Technology, https://doi.org/10.1007/978-981-15-5741-5_9

(Chukyo). Those conurbations, Greater Tokyo and the Keihanshin in particular, boast both some of the highest modal share for public transport in the world and the lowest car usage.

As noted before, Japan's population density as a whole is high. However, Japanese cities are not particularly dense. For instance, Greater Tokyo is less dense than large European metropolis, Greater London[2] in particular which is in turn less dense than most large cities in the continental Europe. The reasons for Japanese reliance in public transport, walking and cycling are other than high density and, as usual, are rooted in history.

Japan was for centuries a feudal state effectively ruled by a military ruler the *shogun*, who receive his authority by delegation from the emperor. In practice, the shogun became a hereditary title. Therefore, we can talk about different shogunate periods depending on the family in charge. Local lords, known as *daimyo*, were feudal subordinates of the shogun who occasionally fought each other and more rarely banded together against their lord. One such upheaval started as a dispute between two daimyos which then escalated to a full civil war that resulted in the end of the Ashikaga shogunate. The aftermath was the Sengoku period, the Age of Warring States, that lasted more than a century. The shogunate was not restored until 1600 when Ieyasu Tokugawa, with the help of Oda Nobunaga and Toyotomi Hideyoshi, unified Japan under a new shogunate led by himself. The Sengoku period was characterised by continuous war. The daimyos' armies were led by samurai, but composed mostly by armed peasants. In order to prevent a repetition to the civil war, the Tokugawa shogunate imposed a new order in which only the samurai were allowed to bear long swords and wear armour so no daimyo leading a peasant army could be able to challenge the shogun anymore. Confucian ideas, in which society was divided into hereditary castes, were strictly followed in the Tokugawa period. Peasants were well appreciated as the providers of food for the empire and considered of a higher status than craftsmen and merchants but attached to the land in a way not too different to European serfs. Thus, Japanese peasants were not allowed to live or travel to village or town other than the one they were born. Agriculture was indeed a successful activity, which, through taxation enriched daimyos and shogun and allowed for an extraordinary increase in population.[3]

On the other hand, daimyos and the samurai at their service were obliged to spend long periods of time in the capital, Edo (Fig. 9.1), and keep a residence there. Given that many samurai also performed most high-level administrative work for the shogunate, Edo became forced residence to up to 400,000 members of the samurai class alone at any given time in the eighteenth century. All in all, counting all the

---

Kawasaki etc., plus huge suburban areas in Chiba and Saitama. Geographers have made some efforts to count the population in the Tokyo employment area. A 2010 estimate put this population on 37 million. We find safe to say that Greater Tokyo has *at least* 37 million inhabitants. Tokyo population estimates are taken from Kanemoto (2015).

[2]Only the Tokyo prefecture, at 7500 inhabitants per square km, is denser than Greater London's 5500. The metropolitan employment area of Tokyo is far less dense at around 3500.

[3]In the seventeenth century alone the population of Japan went from 18 to 30 million with a marked increase in urban population from 1.4 to 5 million.

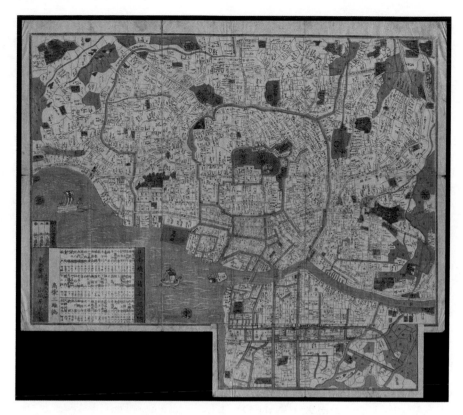

Fig. 9.1 Edo by the 1850s. The commoners concentrate to the south of the shogun's castle on both sides of the Sumida river. Woodcut, image sourced from Wikipedia

servants of the samurai, craftsmen, traders and caste-less workers Edo[4] had about 1 million inhabitants by 1750 and was without doubt the largest city in the world. Osaka, the most important commercial hub at the time, and Kyoto, the official capital and the residence of the emperor were also among the largest cities in the world at the time, with hundreds of thousands of residents. In any case, the cities of the Tokugawa era were rigidly planned, usually around a large castle which served as residence for the local daimyo or the shogun if in Edo. The castle used up a substantial amount of land. For instance, a good part of today's Chiyoda ward, in central Tokyo is occupied by the Imperial palace and surrounding gardens. In Edo times, the shogun's castle grounds were far larger, including the Marunouchi district and Tokyo station. Outside of the castle's moat there were upper samurai residences, built in generous plots of land. Farther away there were the commoner or *mochi* districts were artisans, traders etc. were allowed to live. Their houses were small and set in squared grid.

---

[4]See "The Making of Urban Japan" for a detailed account of the Tokugawa city and its importance in shaping Japanese urban design practice, Sorensen (2005). The same reference is critical for many of our claims about the transformations and adaptations in the Meiji era.

Temples were built in the outskirts of the city, often surrounded by lower samurai neighbourhoods who would garrison outer fortresses in case of need and keep an eye on the lower-class inhabitants of the mochi at all times.

Mochi districts were made of square blocks called *cho*, ideally measuring 109 m × 109 m. Narrow streets of about 4 m wide separate one cho from the next. The cho is divided in rectangular plots were houses stand, but sometimes more squired plots were used to leave space for an internal courtyard. Often those were built up and even narrower lanes built to access them. Although most edification in the mochi districts was just one storey high, they achieved extremely high density, 60,000 inhabitants per square km according to some estimates. That was due to large families sharing a small single or two-room house. By the early nineteenth century the housing in the mochi had improved significantly thanks to the increasing prosperity of craftsmen and traders, the mochi areas expanded and became less dense. At the same time the samurai and daimyo built second residences in the hills. Those were relatively large timber houses surrounded by a nice garden and a tall fence. To this day, this housing type constitutes the aspirational goal of the middle classes.[5]

It is also worth noting the lack of plazas and parks in the traditional Tokugawa city. To some extent, public space was provided in the temple grounds where festivals and fairs took place regularly. A critical difference with pre-industrial Europe is the open space in Japan belongs to the temple, or the castle, is fenced off and can be closed. In contrast, European plazas located in front or just outside a church or a castle are open to the public.

Japan was closed to external influences during the Tokugawa period. Only a few, selected foreign traders were allowed to settle trading posts, for instance the Dejima artificial island in Nagasaki run by the Dutch. Other foreign powers, however, pushed for Japan to open to trade in terms akin to the unequal treaties forced into Imperial China. The best known case is that of the US naval expeditions led by Commodore Perry in 1853 and 1854, who forced the opening of Japanese ports to American trade under threat of bombardment by very destructive state-of-the-art artillery.[6] A rebellion against foreign influence resulted in riots, a civil war the eventual end of the Tokugawa shogunate in 1868. Power returned to the emperor and the daimyos who supported the shogun lost their land, which came under direct ownership of the emperor.[7] That marked the start of Meiji era. The new government was able to import and benefit from foreign practices and technology but was able to retain the reins of power and remain largely politically independent of foreign influence. A period of very rapid economic and population growth started with the Meiji restoration and lasted throughout the Taisho and early Showa eras to the start of WW2. With freedom of movement peasants were attracted to the cities in large numbers. Industrialisation and fast urbanisation followed suit.

---

[5]The aesthetics of those houses came to be understood as Japanese architecture by foreign visitors, see for instance Bruno Taut's admiration of the Katsura imperial villa in Kyoto in Isozaki (2005).

[6]A less known expedition Admiral Putyatin of the Russian Imperial Navy had similar effects.

[7]Thus, becoming available for other uses. That have an impact in the cities.

Fig. 9.2 A view of central Tokyo after the USAAF firebombing on 9–10 March 1945. Sourced from Wikipedia

Natural disasters, such as the Great Kanto earthquake and subsequent fire destroyed much of Tokyo. Most of the housing stock was made of timber at the time, which is a good construction material to withstand frequent earthquakes but remarkably flammable on the other hand. The city was rebuild, mostly using timber, yet again. Tokyo, and many other large cities (besides Kyoto) were yet again thoroughly destroyed by intense fire-bombing by the USAAF in 1945 (Fig. 9.2). Yet, fast urbanisation, the will to adapt to Western culture and the cycles of destruction and reconstruction did not do much to change the urban pattern set in the Tokugawa times. So, how did the large Japanese cities adapt to the new times of fast cultural and technological change?

Back at the start of the Meiji era settlement restrictions related to class were lifted. Now people were allowed to live wherever they could afford a house. That was a boon for rich traders who could now build houses to rival the former samurai or buy from them. Higher rank samurais and especially daimyos now became part of a more or less European style aristocracy who still owned most of the urban and rural land. Much land belonging to the old daimyo residences in central Tokyo was now put to commercial use.[8] In any case, the large surge in population was accommodated in neighbourhoods following the mochi and cho model, with the characteristic 4 m wide streets. Actually, it is not that clear why that happened. An obvious reason is the prevalence of property rights.[9] Rebuilding a very large city with wider streets and larger blocks would require a huge effort in terms of expropriation, compensation and therefore a powerful, economically well-endowed city government. That was far from the case in the aftermath of a crisis that destroyed most of the city. Another reason is technological: craftsmen knew how to build one or two storey timber houses cheaply and effectively. Those houses worked well in the known urban style, so that's the housing style chosen by private developers in the areas around new factories in the late nineteenth century.

An interesting exception to the rule is the Ginza neighbourhood in Chuo, one of the special wards in central Tokyo. Ginza was consumed by a fire in 1872, then rebuilt in Western style. The reason for the Ginza exception could be that of the willingness of performing an experiment, the admiration and willingness to adapt to anything western at the beginning of the Meiji era[10] and the feasibility given by Ginza being a relatively small area. This time it was not the whole city that needed to be rebuilt. New buildings were erected using brick rather than timber and considered fireproof. The whole neighbourhood was planned and designed by British engineers and architects, Colin Alexander McVean and Thomas Waters among them (Sorensen 2005). Western visitors thought the final result very much looked like the outskirts of Chicago or Melbourne. Indeed, the new streets in Ginza were far wider than the customary 4 m, capable to accommodate tram lines and horse-drawn carriages[11] and illuminated by ornate public lights. Other than in Ginza, the majority of wide roads were, and still are, links between neighbourhoods or part of intercity routes, such as the Chuo dori.[12]

As we have explained before, steam railways and trams sprouted all over during the Meiji era. Although foreigners were not allowed to build and own railways in Japan, British technical influence and operational practice was obvious in the early years. The first line, from Tokyo to Yokohama, opened in 1872 with British rolling stock. It employed 300 Britons and other Europeans as technical advisors with the

---

[8]In particular the land owned by disgraced daimyos. Four former residences were used to create European style urban parks: Koishikawa Koraku-en, Rikugi-en, Hama Rikyu Garden and Kyu Shiba Rikuy Garcen. Apart from the Imperial palace gardens most of the other large parks are former temple grounds (i.e. Ueno Park).

[9]See Tiratsoo et al. (2002) and Waswo (2013).

[10]It was very difficult for Jigori Kano, the father of judo, to find anyone interested in teaching him jiu jitsu, the old martial art in which judo is based (Watson 2012). Traditional martial arts were considered a thing of the past, something to be forgotten in the 1870s. Possibly the most famous symbol of westernisation of Japan was the Rokumeikan, a residence for high ranking foreign visitors, famous for parties and alleged debauchery.

[11]At the time most of the commercial traffic was hauled in hand carts to the shops. The canal network was used extensively.

[12]Interestingly only those major roads are named. The myriad of 4 m streets are literally anonymous. Without street names, addresses are complicated requiring the house number within the cho, plus the district, city and prefecture name.

role of educating the local workforce. The track was built to the 3 foot 6 in. Cape gauge (1067 mm), still the most common nowadays in Japan. As Japan lacked coal, electric traction was enthusiastically adopted in suburban lines, which sprouted everywhere around cities once the private main line railways were nationalised in 1906. Those private railways, now often modelled after the American electric interurban standard, played a major role in developing suburban settlements, growing along the tracks and near the stations in land often bought cheaply with the railway before construction. The network run by the national railways also provided service in and around large cities. For instance, the passenger ring railway known today as Yamanote Line exists in pretty much its current form, a double track electric elevated railway, from 1932. This line had tremendous influence in shaping central Tokyo. The Ministry of Railways did not allow the private railways to go beyond the Yamanote Line into the inner districts. That created the need for transfer points that soon developed into commercial hubs known are new urban centres or *fukutoshin*. Most of the inner Tokyo tall buildings, which give the city a reputation for high density, are actually commercial buildings located around the Yamanote Line fukutoshin, i.e. Shinjuku, Ikebukuro, Shibuya etc. Indeed, the majority of Tokyo's population live in individual houses, lining narrow 4 m streets, even in the outer suburbs. Unlike in Europe, there has never been much in terms of rows of apartment buildings in Japan. There are many free-standing residential towers in the suburbs, mostly built after the late 1970s as public housing. It is also important to understand how Tokyo experienced a remarkable population loss to the suburbs starting in the 1970s. By 1995, most of the 23 special wards had lost population relative to 1970, but the prefectures of Saitama, Kanagawa and Chiba went from 7 million in 1955 to more than 20 in 1995. Central Tokyo and the Fukutushin did not lose employment at the time.[13] This population movement, away from work centres, needed a continuous improvement of the rail network. At times the railways could barely cope with the demand, resulting in endemic crowding and slow travel times in the peak. The situation has markedly improved thanks to enormous investments resulting in elevated, four track railways becoming common in both Greater Tokyo and Osaka. At the same time, the metro network in both core cities added several new lines. Often trains travel on a private railway to switch to metro trackage to run inside the Yamanote line perimeter and then switch back to another private railway. This system saves transfers and congestion at terminal stations connecting with the Yamanote line, allows more one sit trips from the suburbs to many areas of the core city and connects outer suburban areas with each other.

The critical question is how the Japanese cities adapted to motorisation in an orderly way. In principal, the post-war governments were not at all opposed to universal car ownership. Indeed, Japan soon became an exporter of affordable, good quality cars to both the West and development countries. There were, however, many obvious problems. Chiefly, the shape and organisation of cities. Tokyo and other large metropolis were rebuilt fast, but for the most part on the previous city grid, which was also and for the most part, used in new developments. It was soon evident

---

[13] See Hirooka (2000) for a details account of the making and evolution of Tokyo's railway network.

Fig. 9.3 Cars off the street in the Japanese suburb. Wikimedia Commons

that such narrow streets would not allow for both parking and circulation. Therefore, the government introduced a national "proof of parking" law by which car buyers need to prove they have an off the street place to park their cars.[14] Another national law forbids daylight street parking in most locations and completely bans overnight parking (Fig. 9.3). Therefore parking, which is excludable and rivalrous and thus a private good, is provided in Japan by the private sector. There are no minimum parking requirements either. The provision of carparks simply follows the rules of supply and demand. Price and availability vary of course a lot between the busy city centres, where some plots of land are dedicated to carparks instead of buildings, to rural areas that are much more car dependent.

   At the same time several pro-motorisation measures were taken. First, the dense on-street tram network inside the Yamanote loop was closed to give more space for internal combustion vehicles. Second, a network of elevated motorways was built in Tokyo on top of the old canals (Fig. 9.5), which became little more than open sewers. Finally, intercity motorways linked the country and road freight traffic mostly replaced rail-based freight.

   In any case, car travel is for the most still not practical in Tokyo, Osaka and other large cities. The existing large road infrastructure, motorways and arterial roads, is congested by commercial traffic. Curiously, the modern substitute of the hand

---

[14]See Barter (2014) for an accurate explanation of street parking rules in Japan written in English.

Fig. 9.4 What we see in this image is common in urban areas in Japan. Pedestrians are supposed to use the sides of the street, and the pedestrian crossing. The lack of car traffic allows them to use the whole width. Note also the mixture of shops and residences and the ugly overhead electrical wires closer to American rather than to European standards. Sourced from Wikimedia Commons

cart used for deliveries and other light cargo is the microvan. A large proportion of motorised transport in Japan's large cities is done with this tiny, easy to park delivery vans. People walk, cycle and for longer intracity trips use the trains. The system of narrow streets in the mochi system facilitates walking and cycling to the station or the local shops. This pedestrian scale development (Fig. 9.4) is facilitated by liberal planning laws approved by the central government and thus valid in the whole country. Japanese planning is characterised by allowing mixed use by default, and by regulating height with a simple formula that limits building height to the width of the street and the setback (Fig. 9.5).

Contemporary Japanese planning is well explained in a 2003 publication by the Ministry of Land, Infrastructure and Transport titled "Urban Land Use Planning System in Japan".[15] It takes just seven pages to provide a fairly detailed summary of the planning system. It is worth to highlight several features. For instance, there are twelve categories of land use. Almost all of them, including the so-called "Category I exclusively low-rise residential zone" allow mixed use. In particular, it permits residential buildings to be used as small shops or offices. Most of other categories add for more uses, for instant larger shops or office buildings. The main exception is

---

[15]Ministry of Land, Infrastructure and Transport (2003).

Fig. 9.5 Nihonbashi, in central Tokyo. The elevated motorway was opened on time for the 1964 Olympic and seen as a sign of modernisation. Sourced from Wikipedia

the "Exclusive industrial zone" which only allows industrial uses. The less restrictive, "Industrial zone" forbids schools, hospitals and hotels, but residences are fine. Shrines, temples, churches and clinics are allowed everywhere, even in "Exclusive industrial zones". The same planning document also details simple rules for floor space ratios and a formula for maximum height. Finally, the national planning law also sets the rules for District Plans and City Master Plans giving some, but not much, leeway to local governments to adapt the national standards to local circumstances, protect historical districts, engage in urban renewal etc.

We believe that the simplicity of Japanese planning laws combined with the limits to car usage are admirable and should become a reference point anywhere in the world. Alas, not everything regarding the built environment is as good in Japan. We have already mentioned the lack of plazas and parks. This is somehow compensated by the amenity of the most common public space, the street, which in Japan is for the most part walkable and interesting, randomly dotted with little shops, restaurants, quirky houses, cats and bicycles, making a stroll always an interesting experience. Open air or covered shopping areas, malls and shopping clusters near railway stations are also vibrant, welcoming and enjoyable even if not strictly public. Finally, temples and shrines provide spaces for much needed contemplation and solace. Much worse than the lack of certain types of public space, or the low quality of the playgrounds, is the fact that most houses are knocked down and rebuilt every 20–30 years.[16] This is a vicious circle; common houses are built to low standards, so nobody wants to buy an old house. Because nobody wants an old house, they don't have resale value and are built to low standards. Only recently the large prefabricated house industry is starting to offer house refurbishment as one of their services. For the most part, a house is knocked down after their occupants leave or die. As a silver lining, real estate speculation is only a thing related to land, not houses, and the prices in Tokyo are stable for decades making it the most affordable metropolis in the world even when its population and GDP keep growing.

Finally, we would like to point out that the success of the Japanese planning is not the result of deliberate decisions but the consequence or inertia. Indeed, we find the results of deliberate planning (Odaiba Island, Minato Mirai 21 etc.) underwhelming in the best-case scenario.

References

Barter, P. (2014). Japan's proof-of-parking rule has an essential twin policy. *Reinventing Parking*. Retrieved March 01, 2020, from https://www.reinventingparking.org/2014/06/japans-proof-of-parking-rule-has.html.

Hirooka, H. (2000). The development of Tokyo's rail network. *Japan Railway & Transport Review, 23*(3), 22–31.

Isozaki, A. (2005). *Katsura imperial villa*. Phaidon Press.

Kanemoto, Y. (2015). *Metropolitan employment area (MEA) data*. Center for Spatial Information Science, The University of Tokyo. Retrieved February 29, 2020, from http://www.csis.u-tokyo.ac.jp/UEA/uea_data_e.htm.

Ministry of Land, Infrastructure and Transport. (2003). *Urban land use planning system in Japan*. City Planning Division, City and Regional Development Bureau, Ministry of Land, Infrastructure and Transport. Retrieved January 02, 2020, from http://www.mlit.go.jp/english/.

Sorensen, A. (2005). *The making of urban Japan: Cities and planning from Edo to the twenty first century*. Routledge.

Tiratsoo, N., Hasegawa, J., Mason, T., & Matsumura, T. (2002). *Urban reconstruction in Britain and Japan 1945–1955 dreams, plans, realities*. University of Luton Press.

Waswo, A. (2013). *Housing in postwar Japan—A social history*. Routledge.

Watson, B. N. (2012). *The father of Judo: A biography of Jigoro Kano*. Trafford Publishing.

---

[16]This rule applies for low quality housing. That's the majority of it.

Chapter 10
Following America, Not Japan: Car Dependent Emerging Megacities

Abstract We describe the problems caused by pollution and congestion externalities in megalopolises such as Jakarta, New Delhi and Kuala Lumpur in comparison with the Japanese experience. We argue that planners in those developing economies decided to spend scarce resources in motorways and freeways rather than public transport, with dire consequences. We also find how some of those cities, many in mainland China, are now developing effective public transport networks with the potential to ameliorate their problems.

Keywords Planning in emerging megalopolises · Congestion · Pedestrian hostility

Mass motorisation reached emerging, rapidly urbanising countries only in the 1990s. For instance, car ownership in Jakarta tripled between 1985 and 2002[1] and it's still increasing at a sustained pace driven by scores of scooter users buying new cars. The story of explosive growth in car ownership and usage is very similar for Kuala Lumpur, New Delhi, Kolkata, Beijing or Shanghai. Notably, none of those cities or countries followed the Japanese approach (Fig. 10.1). Instead of shunning on street parking and fostering active mobility illegal parking is tolerated all over Jakarta where the pedestrian network has decreased in favour of a roads exclusively used by motor vehicles, creating mobility inequality.[2] At the same time, large sums of money are spent in a sprawling network of urban motorways. All the cities mentioned so far have in common tremendous air quality problems leading to an increasing number of deaths related to pulmonary disease.

Mainland China, and to a lesser extent India, have been making impressive efforts to improve urban public transport. For instance, the Beijing subway went from a couple of lines and 434 million boardings in 2000–23 lines and 3.8 *billion* boardings in 2018. With their networks still rapidly expanding, both Beijing and Shanghai

---

[1] Susilo et al. (2007).

[2] Hidayati et al. (2019).

© The Author(s), under exclusive license to Springer Nature Singapore Pte Ltd., part of Springer Nature 2020
P. Guillen and U. Komac, *City Form, Economics and Culture*,
SpringerBriefs in Architectural Design and Technology,
https://doi.org/10.1007/978-981-15-5741-5_10

Fig. 10.1 Bandung, West Java, Indonesia. Just try to cross the road From Wikimedia commons

metros are among the largest best patronised in the world.[3] Planning in mainland China is strongly directed by provincial and city authorities which for the most part favour large, free standing apartment towers separated by some open space and roads with sprawling shopping malls and commercial skyscrapers. In China, private land ownership is not much of a problem so entire, massive neighbourhoods are torn down and built from scratch. New cities for millions are built in a matter of a few years where nothing was before. India, which shares the congestion and pollution problems associated to rapid motorisation in dense urban environments is also expanding its metro networks, especially in New Delhi and Bombay, but the efforts are not yet meeting demand. Note that New Delhi is a master planned city with ample avenues, very sympathetic to car traffic.

Both Jakarta and Kuala Lumpur are finally expanding their public transport network, but again, at a pace too slow compared to China, spending a fraction of the monies going to expand their motorways. Similar problems related to mass motorisation in dense cities can be found in Mexico City, Rio de Janeiro and Sao Paulo in

---

[3]For comparison and according to Wikipedia, the Greater Tokyo rail system clocked almost 15 billion unlinked boardings in 2017. Statistics are available from individual operators, JR East, Tokyo Metro, Toei Subway, Tokyu Corporation and several other private railways.

Brasil (improving), Bogota, Istanbul, Teheran, Cairo etc. If anywhere, those megacities in emerging countries are the ones to be benefit more from the Japanese experience. Alas, one after the other followed the allure of the American private car and motorway route. However, American style suburban growth based on tract housing requires higher expenses in a yet larger network of motorways and arterial roads, plus complex, spread out sanitation. Given that those cities could not afford that, they effectively build dense, chaotic, polluted car-based metropolis. Only when this growth model became glaringly wrong, when congestion bit and stymied economic growth, they started to develop public transport in earnest.[4]

References

Hidayati, I., Yamu, C., & Tan, W. (2019). The emergence of mobility inequality in greater Jakarta, Indonesia: A socio-spatial analysis of path dependencies in transport-land use policies. *Sustainability, 11*(18), 5115.
Susilo, Y. O., Tjoewono, T. B., Santosa, W., Parikesit, D. (2007). A reflection of motorization and public transport in Jakarta metropolitan area. *Journal of the Eastern Asia Society for Transportation Studies, 7,* 299–314.

---

[4] And at a far larger scale than anything imagined in contemporary America. Even Dhaka, in poorest among the poor not longer than two decades ago Bangladesh, is choked by traffic and has a modern metro under construction.

Chapter 11
Motorisation and De-motorisation in Europe

Abstract We describe the European planning approach to motorisation focusing on the historical experience of Barcelona and Paris. That is, we study how relatively large and dense cities had to change to accommodate the motor car and the consequences of such choices in terms of pollution, land use and traffic congestion. We also discuss the European approaches to de-motorisation to find that, in general and in comparison to Japan, they are micromanaged, overregulated and geographically localised in and around the city centres.

Keywords Haussmann's Paris · Plan Cerda · Pedestrianisation · Public transport · Cycle friendliness · Freiburg im Breisgau

When cars became means of mass transportation, politicians in most European made efforts to adapt to the new technology out of fear of losing in terms of economic growth and the favour of the people. There was also strong popular and political pressure in favour of motorisation. However, the both urban shape of the core city and the patterns of suburbanisation are different in Europe than in America or Japan.

We will focus our historical approach to explain the continental European model on the city and metropolitan area of Barcelona as a case study. With more than 5.5 million inhabitants, Barcelona is the third largest urban agglomeration in the EU.[1] Its prosperity has been historically linked to trade. As such, the city experienced rapid economic and population growth when traders based in Barcelona were finally allowed to trade with the Spanish America. At the time Barcelona was under military rule so new construction was forbidden outside the wall perimeter. That caused it to become the densest city in Europe. The military rule and the building restrictions were overturn by mid-nineteenth century. A plan, devised by military engineer Ildefons Cerda (Fig. 11.1), was to guide the expansion of the city, Permanyer and Venteo (2008). At the same time, some urban reform was carried out in the old city, most importantly the demolition of the city walls to create a ring road and allow railway access. Finally, some neighbouring towns, such as Sants and Gracia, were annexed.

---

[1] Fourth if London was to be included, Eurostat (2019).

P. Guillen and U. Komac, *City Form, Economics and Culture*,
SpringerBriefs in Architectural Design and Technology,
https://doi.org/10.1007/978-981-15-5741-5_11

Fig. 11.1 Cerda's Eixample city extension is evident in this picture. Sourced from Wikipedia

The process of extension, reform and annexation was common to any successful city in nineteenth century Europe. City extension was the most important process in Barcelona. Paris, on the other hand, achieved similar results under the direction of Baron Haussmann[2] by opening a vast network of new wide avenues across the old city, *les grands boulevards*, and substituting the decrepit, unhealthy housing stock consisting of two storey houses and small apartment buildings by rows of six-storey plus attic, limestone buildings lining the boulevards.[3] Those buildings are essentially associated with the image we have of Paris nowadays, but at the time of construction derided as monotonous. This type of building combines shops and restaurants, offices and residences often stacked vertically. The success of the Haussmann model coupled with lack of racial tension made European city dwellers to accept living in denser conditions. Before the invention of lifts, many buildings were in fact a cross section of class. The owner would occupy a grand flat just above street level. Middle class tenants would live above, and poorer tenants would live in small rooms in the attic. Lower quality buildings would be occupied by the very poor, often one family per room. In any case, the vast majority of the population would live in flats of apartments. For instance, a good proportion of the *grand bourgeois*

---

[2]Haussmann was a high-ranking public servant, the prefect of the Seine department, working to the orders of Emperor Napoleon III. Haussmann's alma mater was the Paris Conservatoire.

[3]See Carmona (2002) for a detailed account of Haussmann's led urban transformation of Paris.

and the majority of the *petit bourgeois* in Paris always lived in flats (and often paid rent). Those would be more or less grand, large and comfortable, but flats after all. Even in the *Fauburg Saint-Germaine* the so-called *hotel particulare*[4] is more the exception than the rule. Indeed, many buildings in Paris and other continental cities were designed with a *piano nobile*.[5] For example, the narrator of *In Search of Lost Time* lives with his parents in an apartment owned by the Germantes when in Paris. As a child, he does not long for a backyard. He is happy enough to play with his first love, Gilberte, in the Champs Elysees, a public park.

Similar to the Paris Haussmann model, steel frame and stone façade buildings were erected in Barcelona during the second part of the nineteenth century. Those filled the new city blocks in the Cerda's grid pattern. The original plan called for allowing constructions in only two of the four sides of each block, but the municipal government easily bucked to developer's pressure resulting in a very dense city fabric. The new streets and boulevards provided generous space both for pedestrians, half of the total width in Barcelona, and wheeled traffic. Other than for circulation the new streets were supposed to provide aeration, a modicum sun access and allow for better social control[6] than the narrow lanes of the old city. Importantly, the wide avenues of nineteenth century European city very much facilitated the use motor cars when those came up. And they very much took it over. Mass motorisation did not happen in Europe until de 1950s, but the same problems were soon evident. Street parking became scarce and was soon rationed. Illegal parking in squares, unused plots and footpaths became endemic in Barcelona and was not eradicated until the 1980s. Thousands of spaces of underground parking were built by the monopoly SABA (Sociedad de Aparcamientos de Barcelona). The easy-to-drive block pattern of Barcelona was even made more efficient by the early application of centralised traffic light control by local traffic engineer Gabriel Fernandez back in 1958, one of the first implementations of the *green wave* concept in the world.[7] Traffic also grew exponentially in Paris (Fig. 11.2). Being a richer city than Barcelona, it eliminated its tram network as early as the 1930s. The same process did not culminate in the Catalan capital until the early 1970s.

Both Paris and Barcelona, and many other European metropolises, did not lose, but indeed added population during the post war years only to lose some to suburban developments in the 1970s and 80s. Suburbanisation followed a different pattern than in America, where land was cheap and abundant. Both Paris and Barcelona are surrounded by a belt of working-class neighbourhoods made up of large block of flats separated by large carparks, grassland and arterial roads. Local shops are usually concentrated in clusters separated from housing blocks. Large hypermarkets, open

---

[4]An inner-city Parisian detached house, more like a little palace.

[5]That is usually the level immediately above the street, also called *principal* in Barcelona. Sometimes there is a mezzanine (*entresol*) in between the street level and *principal*.

[6]In the form of cavalry charges, if necessary, Carmona (2002). Barcelona lived through countless riots and rebellions in the nineteenth century. That was acknowledged, for instance, by Frederick Engels wrote in 1888 that Barcelona was "the city whose history records more struggle on the barricades than that of any other city in the world", Eaude (2008).

[7]Valenti (2014).

Bundesarchiv, B 145 Bild-F026341-0036
Foto: Gathmann, Jens | 5. Februar 1968

Fig. 11.2 Motorisation in Paris favoured by Haussmann's wide boulevards. Deutsches Bundesarchiv via Wikimedia Commons

air malls with big box retail also abounds in the periphery. This is the European ideal planned suburb, following a strict functional separation that goes back to the work of Le Corbousier and the CIAM (Fig. 11.3). The poor moved out from the city and large wave of immigrants from the provinces came to settle in those developments. The European suburbs are much more car dependent than the city core, which is usually well served by a large metro networks and frequent buses. Still, even nowadays about 20% of intracity trips in Barcelona are motorised, which in such a dense environment causes very serious pollution, accident and spatial problems. The European rich often still leave in the grand nineteenth century apartments in the city. Although there also exist some leafy suburbs composed of single houses, la *banlieue chic* outside Paris.

Other than Paris and Barcelona, any large European city has a form of well patronised rail-based public transport, a metro or Stadtbahn, complemented by buses or trams. This network covers the older, denser parts of the city where walking and cycling are still the preferred modes and that often could not possibly work without public transport. Sometimes a suburban tram, or a commuter or regional railway makes possible to access some outer suburbs. Importantly, public transport was taken over by the city or regional governments early on compared to what happened in the US. Similarly to the US, tram networks disappeared in many countries to leave road space for private cars. This trend was not as uniform as in America, as most cities in central Europe kept their tram systems to the day.

Since the late 1960s and in parallel with the development of European tower-suburbs a trend to limit or even ban car use in the city core gained momentum in

Fig. 11.3 A model of Le Corbousier's Plan Voisin for Paris. The idea was to demolish most of the city and substitute by cross-shaped freestanding skyscrapers. That did not happen, but variations of this model constitute the quintessential Euro suburb. Wikipedia commons

many cities. An interesting early case is that of Amsterdam. Contrary to general perceptions it was in the early 1970s very much full of cars (Oldenziel et al. 2016). The emphasis on public transport, with the revitalisation of the tram network, and more importantly the fostering of a strong cycling culture only started in the 70s. Actually, cycling to work was rather popular all over Europe from the 1890s up until the 1950s. It fell out of fashion when income increased, people could afford cars, and car traffic made cycling dangerous and unpleasant. Cycling is nowadays the main mode of mobility is several middle to large cities such as the already mentioned Amsterdam and Copenhagen.

Private cars still are the main form of mechanised transport in most middle-sized European cities. Such is the case in most Italian cities, with grave consequences, and in places like Klagenfurth, Ljubljana or Vigo where the relatively short distances are not that affected by congestion in terms of travel time. These cities are far denser in the core than their American counterparts, so walking or in some cases cycling is still feasible for short distances. Public transport is normally provided by buses or trams, that are of some quality, more frequent and far more popular than in similarly sized American cities, but relatively low patronised compared to larger European examples. Otherwise, car is often the preferred option for trips to the suburbs and longer intracity travel. Middle size European cities can still function even if eminently car based.

On the other hand, there are quite a few examples of European middle-sized cities that made a conscious effort to minimise, or severely limit, car usage and provide alternatives, i.e. Montpellier, Strasbourg, Freiburg im Breisgau or Pontevedra. The latter is a small city in the Atlantic coast of Spain in which one needs to be ready to walk as driving is mostly forbidden in the inner areas. Montpellier and Strasbourg have modern, efficient and very well patronised tramway networks taking road space from car traffic, which is actively discouraged. Freiburg is possibly the best example of a middle size city making a strong, micro-managed and successful effort to limit car use. Note that the majority of German cities are relatively small and surrounded by a myriad of villages and towns that offer the possibility of living in the countryside. Roads are excellent so it is easy for Germans to find an excuse to drive in and around Heidelberg, for instance. The same model could have worked in Freiburg, but the local authorities decided to use micromanaged policies to greatly limit car usage and foster mobility based on walking, cycling and public transport (see Buehler and Pucher 2011).

Importantly, there is quite a different approach to planning in Europe when compared to the US. State intervention after the end of WW2 in terms of social housing was more common and somehow more successful in Europe than in the US. That's probably because of racism was less prominent in the post-war period, and planning decisions were far more centralised. Planning practice is most of continental Europe is less concerned with separating uses and more with a distinction of rural versus urban land allowing for different densities, heights and floor-space ratios. In particular, there is a lot less emphasis on separating retail and commercial uses from residences. Even if the CIAM principles have resulted in low quality, somehow car dependent suburban accommodation for the poor in Europe those developments are increasingly connected to efficient public transport, and concerted efforts are made to improve their liveability. Improvement and retrofitting is proving not to be an impossible task in any way. In particular, the adoption of Japanese style planning rules seems feasible and, in our opinion, desirable in the European context.

References

Buehler, R., & Pucher, J. (2011). Sustainable transport in Freiburg: Lessons from Germany's environmental capital. *International Journal of Sustainable Transportation, 5*(1), 43–70.

Carmona, M. (2002). *Haussmann: His life and times, and the making of modern Paris*. Chicago: Ivan R Dee.

Eaude, M. (2008). *Catalonia: A cultural history*. Oxford: Oxford University Press.

Eurostat. (2019). *General and regional statistics*. Retrieved February 10, 2020, from https://ec.europa.eu/eurostat/statistics-explained/index.php?title=Regions_and_cities.

Oldenziel, R., Emanuel, M., de la Bruhèze, A. A. A., & Veraart, F. (2016). *Cycling cities: The European experience: Hundred years of policy and practice*. Foundation for the History of Technology.

Permanyer, L., & Venteo, D. (2008). *L'Eixample: 150 anys d'història*. Barcelona: Viena.

Valenti, F. (2014). Los semaforos operativos mas antiguos de Barcelona. *El Tranvia 48 Blog*. Retrieved February 01, 2020, from https://eltranvia48.blogspot.com/2014/07/los-semaforos-operativos-mas-antiguos.html?m=1.

Chapter 12
Conclusions

Abstract We discuss the applicability of the so-called market urbanism based in the Japanese experience to other jurisdictions. We argue that imposing strong limits to on-street parking would be highly beneficial and relatively straightforward in Continental Europe. We review contemporary opinions of commentators regarding some aspects of market urbanism, such as upzoning, mixed zoning and maximum parking rules to American cities. Most public goods in the US are provided and funded locally and, implicitly, planning rules are used to exclude outsiders from the enjoyment of such goods as education, health or leisure. We conclude that market urbanism could only thrive in the US as a result of a deep structural reform and cultural change.

Keywords Market urbanism · Local public goods

We have attempted to provide an economic-planning-culture explanation for different patterns of city growth based on historical facts. We believe, with some caveats, that the management of city growth is better in Japan than anywhere else. Japanese planning is based on few, clear, fair rules that allow for development and at the same time preserve city amenity by keeping most of the private space for the use of pedestrians. Although Japanese cities are notorious for the lack of open public space such as parks and plazas, their absence is more than compensated by a vast, ubiquitous network of pedestrian and bicycle friendly streets and vibrant, interesting, varied shopping areas. We will now discuss whether the Japanese approach to planning and mobility could be adapted to other jurisdictions.

First of all, we think Continental Europe is ready to adapt to the Japanese planning approach, or at least far more ready than the US. Europe would benefit from a more comprehensive approach adaptable to many circumstances rather than the dominant ad hoc, micromanaged planning that produces only pedestrian-friendly areas in the old city areas but leaves the 19th century extensions and the suburbs rather at the mercy of car dominance. It is also true that Japan could benefit from some aspects of European practice in terms of provision (perhaps mostly design) of open public

P. Guillen and U. Komac, *City Form, Economics and Culture*,
SpringerBriefs in Architectural Design and Technology,
https://doi.org/10.1007/978-981-15-5741-5_12

spaces (German vs Japanese playgrounds), see Appendix II for an elaboration on this point with some examples.

It would be very hard to adapt for the US, regardless of the dominant pro market rhetoric. The key is the culture of local (America) vs global (Europe/Japan) public good production. Nolan Grey makes this argument very clear[1]:

> Another reason why Americans might support more restrictive zoning than the Japanese has to do with the peculiar way that we provide public services. Here in the U.S., the quality of essential services like parks, education, and public safety can vary dramatically municipality to municipality. A larger population of low-income households means higher public outlays, higher taxes on middle- and upper-income earners, and lower quality public services.

> Residents respond by self-selecting into the most affluent communities they can afford and pulling up the ladder once they're in. To put it another way, to ensure access to quality schools and open space, American households are encouraged to move out to a suburb and employ land-use regulations to keep out anyone poorer than themselves. This partly explains the extreme racial and economic segregation see in most U.S. cities today [...].

Although the Japanese approach to planning has received, since we started writing this book, some attention from urban economists subscribing to liberal YIMBY (yes in my backyard) ideals, it has also seen by many of the same authors as a pie-in-the-sky. Not only American, but also British scholars have similar opinions[2]:

> And it is those local wishes that are crucial. After all, planning law is ultimately determined by politics. To put it crudely, whether an area has restrictive planning depends on what the electorate wants. Areas that don't want housing will elect representatives to push for tighter planning.

> [...]

> You cannot simply transplant that system to a country with entirely different traditions and socio-political factors. We know of no example of a discretionary system of planning being transformed overnight into a liberal zoning system.

That is, planning culture is captive of tradition and not easy to change, especially in countries with a first past the pole political systems in which each district elects one representative that will be lobbied by local interest if they don't toe the planning line. Those are to different extents the English-speaking countries, and well, curiously Japan. We now know why Japan took their decision in the past, to avoid overwhelming the country with cars parked everywhere and to avoid spending large sums of money changing the shape of their cities. For those reasons, it is very disheartening to observe how the Japanese experience was not taken as a model throughout Asian cities built on similar traditions and throughout the developing world in general where large, messy, polluted and unpleasant megacities are a reality today. Is it too late for them to adapt to a planning culture that works?

---

[1] Gray (2019).
[2] Watling (2019).

References

Gray, N. (2019). Why is Japanese zoning more liberal than US zoning? Market urbanism. https://marketurbanism.com/2019/03/19/why-is-japanese-zoning-more-liberal-than-us-zoning/, Last retrieved on 1 Feb, 2020.

Watling, S. (2019). A Japanese zoning system is no solution to England's housing crisis. Capx. https://capx.co/a-japanese-zoning-system-is-no-solution-to-englands-housing-crisis/, Last retrieved on 1 Feb, 2020.

Appendix A
Public Transport Definitions

Urban railways are usually classified into light and heavy rail. An easy to understand distinction between those two types relies on the breaking distance of rail vehicles or trains. Light rail vehicles (LRVs)[1], or trams, have short breaking distances, so they are capable of street running. They can be driven "at sight" if necessary. That is, drivers are able to keep a safe distance with a preceding rail or road vehicle just by judging a safe breaking distance by sight without the use of specific signal equipment. Light rail usually has level crossings with roads, and those don't require anything other than traffic lights. Note that light rail vehicles can also be utilised in a more or less segregated right-of-way. For instance, tram tracks can be installed in an unpaved reservation running parallel to a road, in a tunnel exclusively used by LRVs main line railways or metros. In that case they will not generally be driven at sight but instead follow appropriate railway signalling and traffic control methods. Heavy rail vehicles are not capable of street running for any practical purposes.

Trams (modern and old), streetcars, Stadtbahn[2] systems and some metros use light rail equipment. In some cases, light rail vehicles use street running in the city centre to switch to main line rail operation to reach further destinations.

Most metros and commuter railways use heavy rail equipment. Metros are mostly underground systems built to serve a dense city core. Sometimes metros are elevated and, in a few cases, they are run at street level. Metros are rarely crossed at level

---

[1] This terminology originated in the US in the early 1970s with the US Standard Light Rail Vehicle manufactured by Boeing Vertol. That was not more than a glorified tram at a time when trams didn't have a good reputation. So, a marketing effort was made to rebrand the mode. It is true that modern trams are often longer and heavier than the ones common everywhere in the early 20th century. However, trams also evolved and became heavier during the 60s in Germany (Duewag) or Czekoslovakia (ČKD Tatra). It is also worth noting that the Boeing LRV was a disaster in terms of reliability.

[2] In the sense of an underground rail network used by trams that in most cases emerge to the surface through portals to run on or parallel to streets. This type of improved tramways were built in several German cities starting in the 1960s. Interestingly, the Tremont Street subway is in essence a Stadtbahn predating the German examples by more than 60 years. Note that both the Berlin and Vienna Stadtbahn networks use heavy rail equipment.

Springer Nature 2020
P. Guillen and U. Komac, *City Form, Economics and Culture*,
SpringerBriefs in Architectural Design and Technology,
https://doi.org/10.1007/978-981-15-5741-5

by roads. For those reasons metros are relatively expensive to build compared to commuter railways or tram systems. They have the highest capacity of all urban public transport in terms of passenger per direction per hour (ppdh). Note that metros can be also extremely successful in less dense urban environments if public transport in the main mode, see the Japanese cases discussed before.

Light rail vehicles running mostly in reserved space in streets are considered a valid, cheaper alternative to metros. They are indeed overall cheaper, although cost may vary widely from one system to other, but offer substantially less capacity. An interesting hybrid is the German Stadtbahn which may run through a tunnel in the city centre and use regular tram tracks once less dense areas are reached.

Commuter railways offer connections from the city centre to suburban areas. Some commuter railways work effectively like a metro in the city centre to branch out when reaching the suburbs. The Stadtschnellbahn is a prime example of this concept (see Berlin).

Bus rapid transit (BRT) is often contemplated as a low-cost alternative to modern tram systems and even heavy rail metro. BRT use road lanes exclusively to offer some of the advantages of modern trams such as metro-like stops and certain independence from road traffic. Additionally, buses can run on the BRT infrastructure and also work as normal buses in other roads. Although BRTs systems offer capacity higher or equal to modern tram systems they come with critical shortcomings. A single bus can only accommodate about 1/3 of the passenger load of a modern tram. To reach the same system capacity a BRT has to run three times more vehicles so a BRT in a dense urban area causes significant pollution problems and acting as a bus motorway in the middle of the city. Note that the busiest BRTs run a bus every 20 s or less! Additionally, BRT stops are often longer, as they usually offer buses to multiple destinations, and critically much wider than tram stops, as they need two wide lanes (one for stopping and another for manoeuvring/passing). The result is a much higher footprint for BRTs than for modern tram systems. BRTs designed for relatively low frequency buses (one every couple of minutes or more) seem a good option for low density areas.

Buses are probably the best option for most suburban areas. They should be replaced by light rail when reaching about 7000 passengers per hour per direction (pphd). The reason is that buses entail negative externalities: noxious fumes and noise. Buses are also harder to electrify or run on batteries as they use a lot more energy per passenger than trams.

Appendix B
Architecture for the Public Space in a Car Free City

We propose changes in planning laws that would result on a marked increase in public transport availability and usage together with people walking or cycling locally or to cover the so-called last mile. All in all, if following the example of Japanese cities, where private car usage does not amount to around 10% of trips. Since this dramatic shift is fundamentally achieved by all but banning on-street parking, there would be interesting consequences for what it can be achieved in terms of public space design. Note that non only more space will be in principle available but also lower pollution and noise, plus an increase in low speed active mobility, pedestrian and bicycles, at street level. An associated change in zoning will result on proliferation of retail and business mixed with residential usage.

Designers can make use of the new available public space and changed conditions. For instance, there will be a new need for bicycle parking. Several bicycles can fit in one car space, so much less space will be needed. Well-designed bicycle parking is a real need and must be taken seriously by architects/designers. The need for space for loading/unloading commercial vehicles, taxis and emergency vehicles to stop and run will not disappear and must also be taken into account. An advantage of restricting the use of private cars by limiting on-street parking is that deliveries and taxis should flow easily. In any case, there would be room to enlarge footpaths in wider roads. Additionally, there is also the possibility of eliminating footpaths altogether is narrow streets. That could be also understood as the whole street being a wall to wall footpath in which cars are allowed. Pedestrians and bicycles can use the width of the roads together with slow running cars. That's public space reclamation. It works in Japan, but the particular designs can be improved.

Finally, there would be more space for trees in city streets. For instance, some streets with low traffic may be somehow too wide. A strip on the centre of the road can be taken over to plant a row of trees.

© The Author(s), under exclusive license to Springer Nature Singapore Pte Ltd., part of Springer Nature 2020
P. Guillen and U. Komac, *City Form, Economics and Culture*,
SpringerBriefs in Architectural Design and Technology,
https://doi.org/10.1007/978-981-15-5741-5

Printed in the United States
By Bookmasters